WORLDWIDE KNOWLE

Ashgate Economic Geography Series

Series Editors:

Michael Taylor, University of Birmingham, UK
Peter Nijkamp, VU University Amsterdam, The Netherlands
Jessie Poon, University at Buffalo-SUNY, USA

Innovative and stimulating, this series enlivens the field of economic geography and regional development, providing key volumes for academic use across a variety of disciplines. Exploring a broad range of interrelated topics, the series enhances our understanding of the dynamics of modern economies in developed and developing countries, as well as the dynamics of transition economies. It embraces both cutting edge research monographs and strongly themed edited volumes, thus offering significant added value to the field and to the individual topics addressed.

Other titles in the series include:

Household Vulnerability and Resilience to Economic Shocks
Findings from Melanesia
Edited by Simon Feeny

Towns in a Rural World
Edited by Teresa de Noronha Vaz, Eveline van Leeuwen and Peter Nijkamp

Global Companies, Local Innovations
Why the Engineering Aspects of Innovation Making Require Co-location
Yasuyuki Motoyama

Economic Spaces of Pastoral Production and Commodity Systems
Markets and Livelihoods
Edited by Jörg Gertel and Richard Le Heron

Worldwide Knowledge?
Global Firms, Local Labour and the Region

MARTINA FUCHS
University of Cologne, Germany

Routledge
Taylor & Francis Group

LONDON AND NEW YORK

First published 2014 by Ashgate Publishing

2 Park Square, Milton Park, Abingdon, Oxfordshire OX14 4RN
711 Third Avenue, New York, NY 10017

Routledge is an imprint of the Taylor & Francis Group, an informa business

First issued in paperback 2018

British Library Cataloguing in Publication Data
A catalogue record for this book is available from the British Library

The Library of Congress has cataloged the printed edition as follows:
Fuchs, Martina.
  Worldwide knowledge? : global firms, local labour and the region / by Martina Fuchs.
    pages cm.—(Ashgate economic geography series)
  Includes bibliographical references and index.
  ISBN 978-1-4724-1016-0 (hardback)   1. Knowledge management.
2. International business enterprises.   3. Labor supply.
4. Globalization. I. Title.
  HD30.2.F83 2015
  338.9'26—dc23

                                                                    2014019270

ISBN: 978-1-4724-1016-0 (hbk)
ISBN: 978-1-138-54657-8 (pbk)

Contents

PART I   INTRODUCTION

PART II   KNOWLEDGE: CONCEPTUAL DELIBERATIONS

PART III   GLOBALISATION OF KNOWLEDGE IN R&D AND PRODUCTION – EMPIRICAL INSIGHTS

List of Figures, Maps, Tables and Boxes

Figures

Maps

Tables

Boxes

Preface

For decades, the academic and public debate surrounding the globalisation of knowledge has focused on scientific-technical knowledge. Today, many contributions clearly illustrate the limits of such a view. This book therefore takes up the challenge of opening up a perspective 'beyond knowledge'. It introduces further kinds of knowledge and interpretation which influence managements' perception of globalisation and therefore the knowledge which is going global.

Thinking and writing about such broad understanding of knowledge is a challenge because of comprehensive, interdisciplinary and long-standing discussions on issues such as knowledge, cognition, perception and truth. Obviously, this book cannot cover the entire world of knowledge. It therefore focuses on economically relevant knowledge, selectively discussing trends out of the complex globalisation dynamics and with particular focus on multinational companies, local labour and regional development.

The book summarises my conceptual thinking and the empirical studies I have conducted since the early 2000s. A sabbatical has allowed me to compile the present synopsis of my work.

This book has benefited greatly from the ongoing cooperation with diverse persons. I have received much support from many academic colleagues who are too numerous to mention here. Particularly, my understanding of knowledge – and the 'interpretation of interpretation' – has been inspired by a joint lecture series for graduate students developed with my colleagues Detlef Buschfeld, Bernadette Dilger, Matthias Pilz and Frank Schulz-Nieswandt. I also drew inspiration from Wolfgang Leidhold and Detlef Fetchenhauer.

Practitioners in companies have also supported this book by sharing their knowledge with me. I thank them for their many important and often unexpected insights on 'worldwide knowledge' today.

The book is based on insights from research projects which were kindly supported by the German Scientific Foundation, the Volkswagen Foundation and the Hans Böckler Foundation.

Especially, I thank Kira Gee for linguistic editing.

The concepts and ideas presented in this publication could not have evolved without these inputs and support. It should be clear, however, that the limitations and imperfections in this book are solely my responsibility.

I dedicate the book to my husband, Christoph Steeg, who supported this project in our daily life. I also dedicate it to Benny the cat who did his best to sabotage it.

Martina Fuchs, Cologne, Germany

List of Abbreviations

BEM	Big emerging markets
CAD	Computer aided design
CENALTEC	Centro de Entrenamiento en Alta Tecnología
CONALEP	Colegio Nacional de Educación Profesional Técnica
EU	European Union
IT	Information technology
LEMA	Large existing market areas
NGO	Non-governmental organisation
OECD	Organisation for Economic Co-operation and Development
PLEMA	Peripheries of large existing market areas
R&D	Research and development

PART I
Introduction

Chapter 1

Worldwide Knowledge? Global Firms, Local Labour and the Region

Many multinational companies have learnt how to globalise knowledge in the last decades. Hence, in multinational companies the 'worldwide knowledge' is changing: managements' subjective knowledge and interpretation of the world – and the knowledge addressed by management. A new worldwide division of labour and knowledge is looming on the horizon. Today, companies in the Global South can no longer simply be characterised as peripheral plants. Yet the upgrading of such firms can take various routes and does not necessarily coincide with regional development. At the same time, the role of headquarters and the positions of the chief production sites in the core economies are also being redefined. From the perspective of regions, new options are emerging for organising the local knowledge base and local learning between companies and other local actors. This is the case not only in the core economies, but also in some peripheries of the world system. Therefore, opportunities are opening up worldwide for sophisticated production and skilled labour, challenging researchers as well as managers, trade unions, workers and other 'regional' actors. Obviously, new conceptual approaches and empirical insights are needed in order to understand these recent dynamics.

This book addresses the following key questions:

- What are the conceptual implications of a comprehensive view of knowledge and interpretation, assuming that both direct managements' decisions with respect to the globalisation of knowledge?
- What are the implications of the internationalisation of scientific-technical knowledge, particularly R&D? Looking beyond R&D, how does production-related knowledge internationalise?
- What are the regional implications of labour dynamics in the Global South, and what do these same dynamics spell for the North?

There are no simple answers to these questions, as they can only be discussed in approximation and from a particular theoretical perspective. The conceptual part acknowledges many valuable contributions on *knowledge* in economic geography and regional sciences, but also in organisational studies, international management studies and the social sciences. Still, the enormous number of recent publications on the topic makes it impossible to cover all facets of knowledge. Hence, the book proposes the 'star of knowledge and interpretation' as a heuristic concept. This reading of knowledge includes scientific-technical knowledge such as research and development (R&D), but also comprises other kinds of interpretation. Furthermore, the 'star' also differentiates between managements' subjective knowledge and 'objectified' knowledge which is the object of globalisation.

The discussion of recent empirical trends in the *globalisation of knowledge* is necessarily limited by the selection of some exemplary trends out of the complex overall dynamics. Using the 'star' as a conceptual lens, the book chooses to focus on manufacturing industries, in particular the metalworking industries, since the scale and duration of internationalisation processes are particularly pronounced in these. Multinational companies have become key performers in the dynamic world economy on account of their geographical scope and importance to different regions. By that token, they have also come to represent crucial driving forces in shaping regional development in different parts of the world, thus connecting the world transnationally. Automotive, electronics, mechanical engineering and various other industries have penetrated foreign countries and integrated new locations into their international networks of production, which over the last decades has created diverse international divisions of labour and knowledge.

Terminological Comments

The book deals with 'firms' or 'companies', with particular focus on multinational companies or corporations. 'Multinational company' is used to denote companies acting at an international level; they include large transnational corporations with strong influence on international and national socio-political arrangements as well as smaller international companies with subsidiaries only in some countries. 'Internationalisation' is understood as roughly comparable to 'globalisation', although globalisation is regarded as more comprehensive and wide-ranging than internationalisation. Particular reference is made to the subsidiaries or affiliates of multinationals. Depending on the context, reference is made to R&D sites and production plants.

The book has a certain bias towards the relationships between the headquarters of multinationals and their foreign subsidiaries. As a result, integration within the company is a key consideration, although relationships between companies such as outsourcing and offshoring are also discussed. Other transactions in global value chains are therefore touched upon.

In the following, the term 'Global South' is largely used instead of 'developing countries' or 'Third World', although none of these terms adequately describe the underprivileged peripheries of the world system. 'Global South' actually suggests a misleading container view of the seemingly poor South and rich North, while 'developing countries' is normative given its implication of economic growth. The 'Third World' has dissolved along with the Cold War, and 'emerging countries' only encompass some of the countries of the Global South.

If featured at all, expressions such as 'developing countries' are therefore only used when quoting specific studies; in these cases they always refer to their particular understanding of the term.

Frequent mention is also made of the 'peripheries of the world economy', in contrast to the mature or advanced cores of the world economy. 'Peripheries' is a functional and relational term which should not be misunderstood in the sense that entire countries are taken to be dependent and poor, a misattribution which is again due to a container view of space. Here, periphery simply means that important companies in the region are in fact subsidiaries, rendering them dependent on foreign external decisions. This setting, which influences regional actor networks and institutional arrangements, is different from that in the core economies. Even though the core economies are also important target regions for foreign direct investments and thus have many foreign subsidiaries, they do tend to host the headquarters of large global corporations and renowned brands, lending them a strong position in the world economy. Given that many regions are situated between the core and periphery, there are also 'semi-peripheries' – a term with a particular connotation in conceptual history (Wallerstein 1979).

In general, notions such as cores, semi-peripheries and peripheries have a Eurocentric or 'triadic' bias (USA, Europe, Japan), as they portray the activities of transnational firms in the core economies via their subsidiaries in the peripheries. Although the following chapters discuss mutual interdependencies rather than unilinear dependencies, this bias cannot be entirely overcome.

Structure of the Book

Chapter 2 elaborates the conceptual framework of economic knowledge in the global economy. Economic knowledge is used as a commodity, as a resource and as capital, an approach which is broadened and specified by the 'star of knowledge and interpretation'.

Chapter 3 begins with scientific-technical knowledge, which is commonly considered a key asset in the knowledge-based society. Since scientific-technical knowledge is often protected by intellectual property rights, the chapter also addresses the contrast between ubiquitous and exclusive knowledge.

The book then takes the important step 'beyond scientific-technical knowledge' (Chapter 4). After some terminological deliberations, knowledge is explained as a social construct, opening up a framework which is then used for discussing informal and implicit knowledge. Experiences are also included here as a relevant concept.

The conceptual design of the book opens up a perspective beyond scientific-technical knowledge and even 'beyond knowledge' itself (Chapter 5). At this point, the concepts of shared visions and patterns of interpretation are suggested as productive theoretical instruments.

Chapter 6 addresses learning, thus taking up a dynamic perspective. It starts with the conception of learning as meaningful action. This view is relevant to understanding the learning of employees in the context of their daily work. Deliberations then shift away from individual to organisational learning. Focusing on the particularities of multinational companies, this part deals with headquarters' problem of 'islands of expertise' in the company's global network of locations. Next, now focusing on international subsidiaries, organisational learning is highlighted as the subsidiaries' localised interaction with multiscalar actor networks and institutional settings in particular regions. Such interaction can lead to upgrading processes.

Chapter 7 discusses spatiality. Regions are a common reference throughout the book, which are defined as the places which bring together the various actors linked in local and multiscalar networks. The practices of actors are shaped by institutional arrangements; vice versa, actors also influence the institutional settings to some degree. With regard to economic knowledge, some places are said to have surplus meaning in the sense of particularly 'creative' or 'innovative' places. Apart from place, space also implies distant interrelationships such as the space of dependency and interdependency.

After an interim conclusion (Chapter 8) the book moves onto capturing knowledge and interpretation empirically (Chapter 9). This part illustrates the methods used in the case studies presented. The case studies are the result of research projects carried out by the author and associated research assistants.

Then, the book gives empirical insights into the globalisation of knowledge in R&D and production. Chapter 10 returns to the 'star of knowledge and interpretation', using it as a framework for empirical perspectives on shared *knowledge and interpretation* (the 'subjective' side of management mindsets) of the globalisation of *knowledge* (the 'objectified' side of globalising knowledge). It should be noted that the book does not refer to the 'globalisation of interpretation'. The portfolio of *managements' knowledge and interpretation of the globalisation of knowledge and interpretation* is therefore left incomplete. This is due to the fact that *globalisation of interpretation* is too broad and vague to adequately capture within the framework of this book. Additionally, it should be noted that the terms 'perception', 'views' or 'interpretation' are occasionally used rather broadly to sum up 'managements' knowledge, shared visions and patterns of interpretation'.

The Chapters 11 and 12 illustrate managements' knowledge and interpretation and the resulting globalisation of knowledge. Chapter 11 discusses the globalisation of scientific-technical, R&D-related knowledge, with a particular focus on global clients and talents. The chapter also highlights the particularities of the product development process and local problem solving as well as the localised interaction with actor networks and institutions to promote R&D. It shows the increasing globalisation of R&D. Using the example of an R&D centre (Delphi in Mexico), it illustrates some indications for change from 'D without R' to co-design of 'R and D' in some peripheries of the world economy. The chapter ends with indications for the internationalisation of R&D-related knowledge into peripheries of the world economy on a micro and macro level.

Chapter 12 discusses the globalisation of production-related knowledge on the shop floor. Like Chapter 11, it acknowledges the importance of global clients and the global labour markets in the perception of management. The particularities of the production process and local problem solving also play a role. Again, as is the case with R&D, localised interaction between the subsidiary, actor networks and institutions at the foreign location is important for the internationalisation of production-related knowledge and plant upgrading. To some degree, labour regulation and labour relations can also work as stimuli for plant upgrading. Thus, in the Global South today, simple assembly plants have frequently taken several steps of upgrading en route to becoming integrated production sites. These trends, visible at a micro level, indicate the internationalisation of production-related knowledge into the peripheries of the world economy. However, little change is still visible at the macro level of nations and regions.

The book ends with an outlook on the impacts on the North (Chapter 13) and a final summary and discussion (Chapter 14).

PART II
Knowledge:
Conceptual Deliberations

Chapter 2
Knowledge in the Global Economy

Asking about 'worldwide knowledge' in the context of global firms, local labour and regional development quickly leads to *economic* knowledge as the key issue and relevant type of knowledge. Economic knowledge is commonly characterised as commodity, resource and capital, in contributions from fields as diverse as economic geography, regional sciences, economics, management studies and social sciences. This chapter takes these notions as a starting point for developing a broader view of economic knowledge, which is expressed below in the 'star of knowledge and interpretation'.

Economic Knowledge

Economic knowledge is knowledge which is bought, sold, transferred and partially stored in various arrangements. It is used as a commodity and today assigned such overall importance that a dedicated, although heterogeneous sector has grown up around it: the sector of knowledge-intensive business services which produces, sells and buys knowledge. This type of knowledge takes many forms, including for example expertise, data bases, or training. A large part of this sector deals with formal, codified knowledge, which is transferable and thus fairly easily transported as a commodity in international trade. However, some knowledge-intensive business services produce highly specific informal, tacit knowledge in close interaction with their clients (Martinez-Fernandez and Miles 2011: 5, Sundbo and Toivonen 2011).

The trade in knowledge is not restricted to knowledge-intensive business services. Manufacturing companies also sell product-related know-how. The machine tool industries for example, or other machine and plant engineering companies, not only sell equipment, but also the know-how required for running and maintaining the machine, usually in close interaction with the client. Thus, knowledge is incorporated into the technology sold, and developed in interaction with the client. The commodity comprises both – the materialised technology and the abstract knowledge required for operating it.

Knowledge as a tradable commodity goes even further. It also relates to incorporated knowledge; that is the knowledge embodied in persons. The human capital approach is an example for such a perspective on economic knowledge, stressing that a well-educated labour force is efficient for the organisation as well as a prerequisite for economic growth (Schilirò 2010). Another example is global talent hunting, which aims to select the best professionals worldwide in order to increase the competitiveness of the company. Recently, the search for the best executives has generated a multimillion US dollar industry of headhunters (Beaverstock, Faulconbridge and Hall 2012: 615). In consequence, talents are considered an important local advantage in regional economies (Karlsson, Johansson and Stough 2012).

Apart from knowledge as a commodity, knowledge as a resource has become a popular view in the last decades in organisational science and economic geography. Inspired by Argyris and Schön (1978), the basic idea is that resources determine the competitive advantage of a company in that the company's specific combination of resources is instrumental in creating and sustaining above-average returns. Often, the resource-based view emphasises the valuable, rare, inimitable and non-substitutable competencies, such as expensive technology, R&D or key patents (Barney 2001).

Whilst the static view of knowledge as a resource sees knowledge as quasi-tangible prerequisite for production processes, the idea of 'dynamic capabilities' acknowledges that knowledge as a resource is subject to learning and thus to change (Eisenhardt and Martin 2000, Teece and Pisano 1994). Knowledge and learning are embedded in contextual and relational settings where the dynamics of evolution and co-evolution often create new results (Bathelt and Glückler 2005: 1545–9). Knowledge is continuously produced and reproduced in interaction. As such, the resource of knowledge differs from material resources because it rarely grows through secret accumulation but through sharing (Ibert and Kujath 2011: 9).

The resource-based tradition has also been advanced with respect to the understanding of standardised versus innovative knowledge. In the early resource-based approaches, routines were the main subject of research. Usually understood as standardised procedures, they are defined as patterns of behaviour and interaction that represent solutions to a particular problem. Routines can be seen as a solidification of practices that develop in order to overcome opaqueness and uncertainties in moments of decision-making (Fuchs and Meyer 2010). Instead of considering them a specific kind of knowledge which could yield insights into knowledge as meaningful action, routines contribute to the stimulus–response-model and the idea of repeatable human behaviour which fits into the technical-organisational procedures of a company. Nelson (2009: 11) for instance explicitly discusses 'routines as technologies and as organisational capabilities'. More recently, the older view on routines has attracted criticism as a rigid approach to describing knowledge; the focus has therefore increasingly shifted to the complex architectures of routines in which superior competencies are guiding the simple routinised behaviour (Amin and Cohendet 2004).

Knowledge is also considered capital. This is true in a direct literal sense since buying knowledge requires money and learning is often an investment. In a broader and conceptual understanding, knowledge can be transformed into other kinds of capital. Bourdieu (1986: 242) talks of 'transubstantiation' of knowledge, describing the continuous re-shaping of knowledge. Bourdieu's distinction between economic, cultural and social capital has recently become very popular, although he states that boundaries between the different types of knowledge are fluid. Bourdieu differentiates between knowledge as economic capital, which can be immediately and directly converted into money and which may be institutionalised in the form of property rights; knowledge as cultural capital, which can be converted into economic capital under certain conditions and may be institutionalised in the form of educational qualifications; and knowledge as social capital, made up of social obligations ('connections'), which can also be converted into economic capital under certain conditions and which may be institutionalised in the form of a title of nobility (Bourdieu 1986: 243).

The views of knowledge as a commodity, resource and capital reflect the general roles of knowledge in the economy. In order to better understand the particular properties of knowledge in the global economy, the 'star of knowledge and interpretation' (Figure 2.1) differentiates between four dimensions, which can be described as follows.

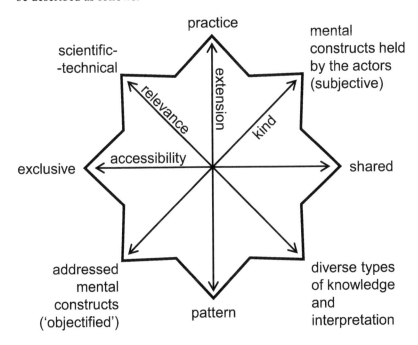

Figure 2.1   The 'star of knowledge and interpretation'

Subjective Knowledge Held by the Actors and Objectified Knowledge

The analysis of knowledge needs to differentiate between subjective and objectified knowledge. 'Subjective' knowledge describes knowledge held by the actors themselves. An example is the subjective knowledge shared by a global company's top managers, which is accompanied by and embedded in interpretations and which direct managements' decision-making. But there is also 'objectified' knowledge, which is knowledge addressed by the actors, such as the knowledge which is spread to different plants in order to enable global production. With regard to worldwide knowledge and global firms, subjective and objectified knowledge are two sides of the same coin since economic decisions and practices clearly include both: the acting subjects on the one hand (with their held knowledge) and the object addressed.

Knowledge as Practice and Pattern

The acquisition, adoption and appropriation of knowledge – learning – are practices. Learning largely addresses knowledge as patterns, defined as commonly shared and rather durable mental constructs (for example learning the grammar system of a language). Apart from knowledge, learning also addresses shared visions and patterns of interpretation (for example, learning how to behave correctly in a contemporary company organisation).

Exclusive and Shared Knowledge

Although knowledge can be widespread, nearly ubiquitous and universally accessible, it is often highly exclusive to companies, workers and regions. In many firms, R&D in particular is highly exclusive; companies closely guard patents and other intellectual property rights in order to retain their competitive edge in the product markets. Knowledge is also often exclusive with regard to labour as only some workers have access to training programmes that increase their competitiveness on the labour market.

Scientific-Technical Knowledge and Diverse Types of Knowledge and Interpretation

Scientific-technical knowledge is relevant economic knowledge, but it is not the only form of knowledge or mental construct required by companies, employees and the knowledge-based economies. There are additional kinds of knowledge and interpretation which extend 'beyond' scientific-technical knowledge. In the discussion on economically relevant knowledge, such other concepts are often ignored. Here it is argued that situational 'shared visions' and widespread 'patterns of interpretation' are closely related to knowledge and as such influence economic decisions, strategies and action in place and space alongside knowledge. Hence, a perspective is opened up on mental constructs 'beyond knowledge'.

The star of knowledge and interpretation represents the conceptual framework upon which this book is based. The next chapter begins by looking at scientific-technical knowledge, which is the starting point of the argument. Economic knowledge is then differentiated further and integrated into a broader framework of interpretation, acknowledging that economic decision-making and practices are influenced by a variety of mental constructs.

Chapter 3

Scientific-Technical Knowledge and the Knowledge-Based Society

Knowledge has always been a driver of economic growth and social development. The discovery of fire was clearly influential, as was knowledge about hunting, farming and the process of smelting metals (Cooke 2009: 69). Modern scientific-technical knowledge can be traced back to ancient philosophies and sciences, the Age of Enlightenment and industrialisation.

Today, scientific-technical knowledge generated by R&D is often put on a level with knowledge per se. Indeed, scientific-technical knowledge is an important cognitive concept and way of perceiving the world, be it for companies or knowledge-based societies. Scientific-technical knowledge contributes to understanding the world, is essential for survival, and helps to change and control the world, or at least parts of it. Many insights gained through scientific-technical knowledge are exciting and thrilling, in particular *scientific* knowledge which is often spectacular. Scientific knowledge helps to understand the position of humans in the universe as part of physical matter and in the context of geological time and biological evolution. The same is true for *technical*, applied knowledge. Apart from being practical and useful, technical knowledge can also be fascinating and inspiring, for example the ambitious architecture of new buildings, the speed of cars, trains and aeroplanes, or new applications in electronic communication.

The OECD definition of R&D reflects this affirmative understanding of scientific-technical knowledge. R&D is regarded as

> creative work undertaken on a systematic basis in order to increase the stock of knowledge (including knowledge of man, culture and society) and the use of this knowledge to devise new applications. R&D covers three activities: basic research, applied research, and experimental development. (OECD 2013a)

From this perspective, scientific-technical knowledge is an integral part of economic, social and cultural development, such as the knowledge embodied in medicine or environmental technologies. Scientific-technical knowledge produced by R&D offers new insights and applications and new knowledge proven by experiments.

But scientific-technical knowledge – with its high renown – also brings with it considerable problems. In an era of worldwide competition, global society increasingly depends on scientific-technical knowledge. The more scientific-technical knowledge is produced and used, the higher the attendant risks. There are increasing problems of controlling scientific-technical knowledge such as nuclear energy, biotechnology and genetics; resolving such control deficiencies requires yet more scientific-technical knowledge. Hence, the increase of scientific-technical knowledge leads to opaqueness and cumulative problems of risk control. Paradoxically, scientific-technical knowledge leads to greater ignorance and lack of knowledge: the more we know, the greater our awareness of not knowing; the more we use scientific-technical development for new technology, the larger the areas of ignorance of how to control the risks inherent in such technologies (Willke 2001, Howells 2012). As a result of these knowledge and control deficits new scientific-technical knowledge keeps becoming ever more necessary and important to managers, politicians, planners and people generally.

Scientific-technical knowledge is an important facet of knowledge because of its obvious relevance to global society. At the same time, it is dangerous to overemphasise this type of knowledge and consider it the *only* kind of knowledge.

Scientific-technical knowledge (as physics, chemistry, biology and engineering) is produced by specific methods based on an epistemology with strong roots in positivism, critical realism and critical rationalism. To a large degree, economics and – to a slightly lesser extent – social sciences use a similar epistemology to produce 'rational' knowledge. Such disciplines and sub-disciplines form part of the naturalist tradition of investigation which considers nature as independent of the researcher and the research context. Rational knowledge is formalised, specialised, representative, replicable, calculable and as objective as possible. It is generated under controlled conditions, often in laboratories or test centres, and follows clear methodological standards. For the respective strands of quantitative economics, the guiding paradigms are rationality and *ceteris paribus* assumptions. Furthermore, such knowledge is institutionalised in research centres and departments and generated by professionals, often in networks of experts. It is monitored and evaluated, partially by other scientists and partially by consultants and auditing agencies (Knoben and Oerlemans 2012).

Scientific-technical and 'rational' knowledge as used in economics are closely related. The general Walrasian economic equilibrium theory, which is based on rational choice, assumes information to be complete, perfect and the same for all agents. The Walrasian approach tends to focus on knowledge in the sense of 'data' and 'information' since the notion of contextualised and partially embodied knowledge is too unwieldy for modelling.

Here, in the scientific-technical paradigm, with its epistemology of positivism and rationalism, 'data' are concrete elements of subject matter or bits of informational content. Information is derived from data, and information can be contextualised into knowledge (Dühr and Müller 2012: 423).

In this understanding, data and information can exist without interpretive knowledge. If a chimpanzee – unable to read the human code – plays with a book, the data and information contained in the book will survive the game as long as the chimpanzee does not destroy or eat the book (Dosi 2012: 171). From that perspective, data and information form part of the structural sphere, which survives for a longer time span.

Viewed differently, one might argue that data and information only exist when they are perceived and thus put in an interpretive social context – in this example, the context of persons able to read the book. Thus, one could state that knowledge only exists in the context of subjects familiar with the rules of interpretation, subjects thus who are able to understand the contents (Table 3.1). Current approaches in game theory, innovation theory, new growth theory, organisational studies, spatial economics and microeconomic theory are now contributing to a more complex understanding of the contextual embeddedness of knowledge (Arena, Festré and Lazaric 2012: 1–2). Chapter 5 will return to the topic of rules of interpretation and discuss this topic comprehensively.

Table 3.1   Data, information and knowledge

	Data	**Information**	**Knowledge**
Characterisation	Bits of informational contents with a low level of abstraction	Perceived and codified contents derived from data	Contextualised information, and contents and modes of understanding it
Example	The letter sequence b-o-o-k	Content of a book, such as a dictionary of words	The contents (translation of the words) and the heuristics to read, understand and interpret the words
Rules of understanding	Do not contain the rules of how to understand the data per se	Does not contain the rules of how to understand the information per se	Includes the rules of understanding

In large parts of society, there is still a tendency to overestimate scientific-technical findings. Because of its formalised, representative, calculable and objective nature, scientific-technical knowledge is often assigned independent facticity and universal validity. Data and information are – quite incorrectly – simply considered 'facts'. In consequence, there is a tendency towards the hypostatisation and reification of scientific-technical knowledge as an autonomous entity and material substance. Seen from this perspective, scientific-technical knowledge accumulates, inevitably progressing towards a larger body of knowledge and also leading to 'higher' knowledge (Ibert 2007). But although it can be accumulated, scientific-technical knowledge is not a hard and tangible substance – it is a mental construct. The hypostatisation and reification of scientific-technical knowledge implies that experience-based knowledge, shared visions and patterns of interpretation are weak in comparison and somehow less important, prone to fading away even though they play an important part in the globalised economy. The view of scientific-technical knowledge as solid, tangible, universal and everlasting is illusive. It is shaped by biological and cultural evolution in that it depends on human senses, presuppositions, the (limited) perception of scientists, language and state of the art research methods. Scientific-technical knowledge is only valid in the context of time and space and in social and cultural contexts. As such, scientific-technical knowledge is subject to change, sometimes changing gradually and continuously and sometimes profoundly as part of scientific revolutions (Kuhn 1962/2012). Knowledge generally is deeply rooted in and framed by the interpretations, ideas and belief systems prevalent in a particular society at any particular time. Even 'hard facts' come under fire and fade away. The history of science offers many examples of how supposedly 'solid' knowledge became fragile and then dissolved.

One conclusion is thus that scientific-technical knowledge has different facets. Scientific-technical knowledge leads to high technology, bringing advantages to companies in global competition. At a more general level, it is central to explaining the world, creating benefits to individuals and societies. At the same time, scientific-technical knowledge can be risky if overvalued. Other cognitive constructs exist which are not rooted in the scientific-technical paradigm. Although the 'solid' sciences are held in high esteem by society, these other types of knowledge can be just as important for companies, workers and regions. Experiences are one such example. Low-tech production is another which is a vital imperative for many companies and their employees. Before considering this in more detail the next section discusses knowledge in the context of the knowledge-based society. Interestingly, here the debate has already shifted away from a near-exclusive view on scientific-technical knowledge towards a broader understanding of knowledge.

The Knowledge-Based Society

The discussion of the knowledge-based society is rooted in the diffusion of information and communication technologies in the 'post-industrial society' (Bell 1973) and the debate on sectoral economic change, where a shift was observed away from primary raw material and industrial production towards a service-based economy (Thompson and Harley 2012: 1369). The increasing focus on the knowledge-based society coincided with the race to the stars during the Cold War, reaching a peak after the lunar landing of the Soviet Union and the shock this implied for the USA and Western Europe. Scientific-technical knowledge has been viewed as an opportunity for companies or nations in meeting the challenges of competition ever since. Today, knowledge is considered the most important impetus for innovation in economy, going along with the scientification of society, in which science and its technological artefacts penetrate almost all societal parts (Adam and Westlund 2013: 1–2).

The knowledge-based society has since become synonymous with a new period of economic and social development. This notion presents a range of fundamental questions. Objections refer to the normative implications and the compatibility of knowledge with other drivers of economic growth. For example, overemphasis of knowledge production ignores the importance of manufacture of material goods and manual work (Leydesdorff 2006: 15) – a topic which was heatedly discussed during and after the property crisis and the 2008/2009 economic recession.

Another problem is the key role ascribed to scientific-technical, rational knowledge in effecting social change. Schütz (1943) was one of the first to discuss 'the problem of rationality in the social world', with Hayek (1945) raising the fundamental question of the importance of scientific and rational knowledge (see Arena, Festré and Lazaric 2012: 1). Even though Hayek was far from denying the usefulness of equilibrium analysis (Hayek 1945: 530), he described its relevance as follows:

> ... one kind of knowledge, namely, scientific knowledge, occupies now so prominent a place in public imagination that we tend to forget that it is not the only kind that is relevant. ... Today it is almost heresy to suggest that scientific knowledge is not the sum of all knowledge. But a little reflection will show that there is beyond question a body of very important but unorganized knowledge which cannot possibly be called scientific in the sense of knowledge of general rules: the knowledge of particular circumstances of time and place. (Hayek 1945: 521)

Recently, revised versions of the knowledge-based society and knowledge economy have become fairly fashionable, leading recent economic policy analysis and debates in the Organisation for Economic Co-operation and Development (OECD 1996) and in the European Union (Westeren 2012: 1). Often, these concepts go beyond simple notions of the 'information age' or the simplistic belief in scientific 'facts'. Current academic contributions stress that different kinds of new knowledge are important to the economy and society.

Gibbons (2003: 231) distinguishes Mode 1 and Mode 2 as two modes of knowledge production. In Mode 1, problems are resolved mostly in the academic community, in Mode 2 knowledge is produced in a context of application. Mode 1 is considered disciplinary, Mode 2 interdisciplinary. Mode 1 has relatively homogenous skills, Mode 2 more heterogeneous skills. Mode 1 is hierarchical, Mode 2 less vertically ordered. Mode 2 also includes more informal knowledge and shows more reflective types of knowledge and interpretation, such as responsibility.

A similar typology is used in relation to Fordism and post-Fordism (Textbox 3.1). Rutten and Boekema (2012: 984–5) regard the knowledge society as the new, post-Fordist period which followed on from the era of Fordist mass production. The new period, which is characterised by a particular type of knowledge production, transformation and use, can be separated into two stages. The first stage is the knowledge economy 1.0, with flexible specialisation and innovation particularly

Box 3.1   Knowledge and learning in the transformation from Fordism to post-Fordism

---

Fordism was first brought into the discussion by the French regulation school (Aglietta 1976, Boyer 1987, 1995). As a theoretical approach it seeks to explain socio-economic change by focusing on the transformation of different institutional arrangements. Institutional arrangements are produced and reproduced by powerful actors such as governments, employers' associations, trade unions, non-governmental organisations (NGOs) and other interest groups; they rule labour relations. In capitalist societies, labour relations show fissures between the spheres of production and reproduction, but tensions are overlaid by social compromises which for the most part are set down in institutional regulations. Social transformation sets in when the fractures and strains become too strong. In the time after the Second World War for example, the core world economies and their market-based, competitive societies were able to change themselves by self-transformation (Jessop and Sum 2006, Peck 1996).

A well-known example for transformation is the change from Fordism to post-Fordism. Fordism was dominant in the USA and Europe and refers to Henry Ford's introduction of mass production and affordable products for employees (in Ford's case cars). Based on the technology of the assembly belt, Fordist production was extremely specialised and centered on high division of labour. Competencies were centralised at the top of the company hierarchy; blue-collar workers on the assembly line only needed brief on-the-job training. Henry Ford implemented Frederic Winslow Taylor's (1911) ideas of 'The principles of scientific management'.

During the 1970s and 1980s, problems such as market saturation, new constellations on the world markets and bureaucracy in the large hierarchies of companies and state organisations induced the change to post-Fordism. In factories, flexible specialisation and economies of scope began to replace mass production and economies of scale. Decentralisation of knowledge became more important for production and was implemented via training, group work and participation in decision making. Motivation replaced direct control at least partially. Participatory knowledge is considered fundamental in effecting this change.

Today, significant new and diverse trends indicate there is no purely post-Fordist production. Different varieties of capitalism are interwoven at a global level. Even within a company or production plant, mixed systems and diverse recombinations of rigid and flexible production are encountered. However, it is evident that modes of production and thus the knowledge held by workers have changed profoundly in the last decades. The notion of 'industrial transition' refers to the open process of change and its different paths, with strong emphasis on globalisation. It thus refers to the global-local interplay in manufacturing, which includes the economy, ecology and society and creates new requirements for innovative knowledge (Fuchs and Fromhold-Eisebith 2012: 233, 237).

in manufacturing industries. The authors refer to this as the 'second industrial divide', first mentioned by Piore and Sabel (1984). The second (contemporary) stage is the era of the knowledge economy 2.0, which is based on global web-based communication devices such as smart phones and tablet computers. Innovations driving the socio-economic system are not primarily concentrated in manufacturing but rather in loose social and professional networks of individuals (Rutten and Boekema 2012).

Today participation is considered a central element of the knowledge-based society. Hinchliffe (2000: 581) describes the change from traditional management control to a new kind of management. In his view, traditional management was based on orders and strict rules, a clear division of tasks, strongly routinised work, the importance of knowing and keeping one's place, and a clear division of rationality (work) and emotion (leisure). In contrast, the new managerial style is characterised by flexibility, creativity, self-organising units or teams, semi-autonomous elements, encouragement of change, openness to criticism, and work as life and leisure in a 're-enchanted world'. Leydesdorff (2006: 15) goes as far as recognising new decentralised, networked governance mechanisms which exist alongside market economy and governmental power.

The notions of 'knowledge created by participation' are essentially heuristic concepts which help to understand change in knowledge production. However, such concepts only refer to selected visible trends. Focusing on the promising new world of human labour, there is a tendency to forget the constraints imposed by global competition and to ignore those who work in the shadow. For many employees, the 'creative, self-organising, re-enchanted world' with flat organisational structures described by Hinchliffe (2000: 581) is a utopian dream, regardless of whether they work in the core economies or in the peripheries of the world, and regardless of their status as blue- or white-collar workers. The knowledge-based society does not release human labour from the need to sell itself on the labour market. Dangerous and physically demanding work is still widespread. Work is still exhausting even in the advanced manufacturing industries. The digitally controlled, fully automatic, computer-guided and deserted factory foreseen in the 1980s has not materialised. Assembly continues to be largely monotonous in the core economies and peripheries of the world system alike, and dirty and hard manual labour still exists in the Global South as well as in the North.

In the highly regulated core economies, workers and their representatives have partially succeeded in their struggles for improved working conditions and safety at the workplace. During the last decade some such improvements have been 'exported' to the Global South, for example as general standards implemented or specifically as part of technology transferred. International agreements and the commitment of national and international trade unions and NGOs have also contributed to improving working conditions in the factories of the Global South (see Chapter 12). Nevertheless, in many such countries where a large part of production for the world market takes place, work is still unsafe, burdensome, and poorly paid.

Additionally, decentralised, web-based communication which brings people together from different parts of the world – as suggested by Gibbons (2003: 231) and Rutten and Boekema (2012: 984–5) – is only accessible to particular workers, people and regions. Exclusiveness and accessibility to knowledge are general problems in the worldwide distribution of knowledge which is highlighted by the following chapter.

Ubiquitous and Exclusive Knowledge

Economists divide commodities into public or private. The ideal public commodity is characterised by non-rival use, universal access and lack of exclusive appropriation, whilst the opposite applies to a private good. In empirical cases, however, knowledge as a commodity is a composite of rivalry and non-rivalry as well as exclusivity and broad access (Witt, Broekel and Brenner 2012: 375–6). In the case of economic knowledge, and under the conditions of global competition, rivalry and restricted access to scientific-technical knowledge tend to be the norm.

Exclusive knowledge – such as patents, copyrights and trademarks designed to protect intellectual property – prevents the transfer, mobility and migration and thus the 'ubiquification' of knowledge (Malmberg and Maskell 1999). Despite this, competitors imitate innovative designs and construction plans, harming the companies that have invested much in developing the innovation. Given the race for scientific-technical knowledge and innovation, it is hard to keep knowledge in place for any extended period of time. Legal and illegal imitation have been used as a means of advancing industrial development since the beginning of the industrial revolution, pointing to the difficulties of encapsulating knowledge despite the existence of patents, copyrights and trademarks. This is especially true for technology-intensive companies. Permanent innovation and learning are therefore required in order to keep one step ahead of the global competition (Spencer 2011: 48).

From the perspective of the technologically less developed companies, nations and regional actor networks, exclusive knowledge and intellectual property rights tend to be acknowledged as a commodity of the rich and powerful. They therefore question such rights of disposal. A prominent topic in the media is the secrecy of knowledge held by pharmaceutical companies. Political activists supporting the Global South argue the illegitimate nature of profit-making by such companies based on the patents they hold for medical drugs. From this point of view, globalisation as practiced by multinational companies as knowledge holders spells the loss of control for the underprivileged of the world economy (Mittelman 2004: 15). Hudson (1999) illustrates the situation as follows:

> Successful firms and regions thus guard it [the knowledge] jealously. If, however, firms 'learn' via producing and protecting such knowledge, if 'regions' seek to learn in the same way, in the final analysis this is to enhance their

competitiveness in a range of markets. … Knowledge and learning may be necessary for economic success but they are by no means sufficient to ensure it; nor, even more so, are they sufficient to ensure equality, cohesion and social justice. (Hudson 1999: 59)

Thus, the debate surrounding the knowledge held by companies and regions is just a new twist in the well-known insight that 'knowledge is power' (Hudson 1999: 59). New political formations have recently developed called the 'access to knowledge movement', or 'A2K' (Krikorian and Kapczynski 2010: 9). In some countries such as Sweden, Austria and Germany, the *Pirates* have become established as a new political party whose key issues are globalised digital knowledge and a new order of access to knowledge. Thus, scientific-technical knowledge is an important feature with regard to power and dependencies, in particular from a global perspective.

Scientific-technical knowledge can thus be described as segmented socio-economically and geographically. Even though there are opportunities for some employees in certain companies and in some regions to progress, in other places the possibilities of employees are severely limited. Cumulative effects often apply, supporting those who are already better off and adversely affecting those who are not. Cooke (2009: 68) draws on the biblical 'Matthew principle' of 'to those that have, more shall be given' and Veblen's remarks on cumulative causation (Veblen 1898: 384–94). Myrdal (1957: 4) pointed to the growing worldwide disparities between under-developed and highly developed countries, stating that for the latter, 'all indices point steadily upwards. On the average and in the longer span there are no signs of a slackening of the momentum of economic development in those countries'. He then goes on to develop the well-known principle of circular and cumulative causation (Mydral 1957: 11–20).

However, there is neither a singular, clean and plain trend of downgrading, nor an instance of cyclical polarisation or 'polarisation reversal'. Sometimes, there are companies with clearly structured knowledge-intensive production and centralised competencies in the North and peripheral extended work benches in the Global South. Theoretical models such as the 'regionalised product life cycle' describe the principle of high-tech product innovation in the core regions of the world economy and low-tech in the peripheries of the Global South (Michie 2011: 12–13). At the same time, significant upgrading processes and learning occur in peripheral production plants and regions. There is no one-way street towards the uneven division of knowledge. At the same time, there is no clear indication that the disparities with respect to knowledge are decreasing and that there is a quasi-automatic trend of convergence, driven for example by new information technologies and the web-based media. The empirical reality in the global production system is manifold, inconsistent and contradictory (Chapters 11–13).

.

Chapter 4
Beyond Scientific-Technical Knowledge

The last chapters have illustrated that scientific-technical knowledge plays an important part in determining power and dependency on a worldwide scale. Clearly, though, scientific-technical knowledge is related to other forms of knowledge and interpretation. The following chapters broaden the focus beyond scientific-technical knowledge to include topics which have so far only been touched upon.

This chapter begins with the basic tenet of 'social construction of reality' (Berger and Luckmann 1966/1991). Additional kinds of knowledge are then brought into focus, such as informal and implicit knowledge and, in particular, experiences. The perspective is then extended even further, not only *beyond scientific-technical* knowledge, but *beyond knowledge* itself (Chapter 5). This is where other forms of interpretation – especially shared visions and patterns of interpretation – come into play.

Beyond Scientific-Technical Knowledge: Knowledge as a Social Construction of Reality

The sociology of knowledge discusses knowledge as a product of society and thus framed by social settings in historical time and in particular spaces (Scheler 1925/1960, Mannheim 1925). The idea became especially popular with the publication of Berger and Luckmann's (1966/1991): *The Social Construction of Reality: A Treatise in the Sociology of Knowledge.*

Berger and Luckmann (1966/1991: 13) regard knowledge as socially constructed during the process of communication. Although this puts knowledge on a par with other social constructs such as interpretation, belief, or opinion, they stress that knowledge is used in communication in a particular way. Knowledge is set apart by the shared conviction that phenomena of the environment are real and true. It also meets additional specific conditions such as relevance, explanatory power, evidence, appropriateness and intersubjective confirmation (Berger and Luckmann 1966/1991: 13). Knowledge thus becomes 'justified true belief' rather than simple belief without validation. As Berger and Luckmann (1966/1991) explain at the outset of their book:

> The basic contentions of this book are implicit in its title and sub-title, namely, that reality is socially constructed and that the sociology of knowledge must analyse the process in which this occurs. The key terms are 'reality' and 'knowledge', terms that are not only current in everyday speech, but that have

behind them a long history of philosophical inquiry. ... [We] define 'reality' as a quality appertaining to phenomena that we recognize as having a being independent of our own volition (we cannot 'wish them away'), and ... define 'knowledge' as a certainty that phenomena are real and that they possess specific characteristics. In this ... sense ... the terms have relevance both to the man in the street and the philosopher. (Berger and Luckmann 1966/1991: 13)

Since then, different perspectives of social constructivism have developed, embedded in traditions of hermeneutics and the interpretive paradigm and employing diverse methods of qualitative research (Flick 2009: 69). They share the basic tenet that reality cannot be experienced directly and independently, but is constructed by the perception of signs and the interpretation of their meaning. This is true for all kinds of knowledge including scientific-technical knowledge, which is a product of economic, socio-cultural and political conditions in time and space. Scientific-technical knowledge is continuously expanded, changed, renewed and reframed; as such it is part of the society which produces it and influences that society in turn.

Based on these assumptions, Knorr-Cetina (1999) argues that the scientific-technical knowledge produced in research laboratories and test centres is not 'objective', but an outcome of 'epistemic cultures'. As a result, science has to be 'put in its place', as postulated by Livingstone (2003) in the title of his book. However, positivist, objectivist and rationalist scientists only partially accept the place-dependency of science. According to Shapin (1998: 5), modern scientists do tend to acknowledge the influence of personal preferences and cultural contexts on the production of scientific ideas; at the same time many scientists share the conviction that scientific modelling itself is a matter of context-free conditions. In contrast, the qualitative approaches employed in the social sciences regard the entire process of scientific enquiry as socially constructed, including empirical research.

Even though the 'social construction of reality' has been a long-standing code for epistemological commitment and confession (Hacking 1999: vii), and even though in economic geography the culturally inspired views of economics were regarded as a Pandora's Box (Thrift 2000: 689), today's research community tends to accept such insights and implements epistemological notions in its methodology. Knowledge is not true per se; knowledge is true only for the members of the society or community who consider it to be true. Based on this fundamental understanding, the following section refers to mental concepts beyond (formal and explicit) scientific-technical knowledge.

Beyond Scientific-Technical Knowledge: Informal Knowledge, Implicit Knowledge and Experiences

Today, issues 'beyond knowledge' are suggested as the new fundamentals in the analysis of knowledge. The growing interest in a broader understanding of knowledge is in line with notions of the 'knowledge economy 2.0' or 'Mode 2', which highlight the informal knowledge generated in vertical communication structures and in loose networks, thus overcoming the boundaries of academic disciplines and borders between countries (Gibbons 2003: 231, Rutten and Boekema 2012: 982–5). At the same time, this broader understanding of knowledge leads to a certain opaqueness since diverse kinds of non-formalised knowledge are often mentioned in the same breath.

Addressing informal and implicit knowledge, this section therefore highlights specific features emphasised by the current debate as relevant knowledge. Experiences have become rather important to the recent discussion of knowledge in production and working life, which is why they are then described in more detail as a particular form of contextual knowledge. The discussion is then extended to encompass an even wider range of mental constructs.

Informal Knowledge

Distinct from formal scientific-technical knowledge, non-formalised or informal knowledge is often viewed as not representative, not (easily) replicable and incalculable. It is not generated under controlled conditions, does not follow strict methodological standards like the production of formal knowledge and is not institutionalised in research centres and departments.

Asheim and Coenen (2007: 661–4) and Asheim (2012: 997) have attempted to improve the conceptual distinction between formal scientific-technical knowledge and informal knowledge. They distinguish between 'analytical' and rather formalised science-based knowledge and 'synthetic' engineering-based knowledge, which stands between formal and informal knowledge. Synthetic knowledge is applied, problem-solving, interactive, partially codified and tacit, context-specific and rather dependent on particular situations. The third category is 'symbolic' knowledge, which is rather informal, arts-based and highly variable. Asheim and Coenen's notion of analytical, synthetic and symbolic knowledge has become quite popular (see Gertler 2008), although problems persist with the terminology and typecasting (Martin 2012a: 31).

Recent strands of literature stress that informal knowledge is often generated via 'buzz'. Rather than referring to a particular kind of knowledge with specific attributes, buzz explains how such knowledge is generated. The term refers to the 'buzzing' sound of insects and thus to the concept of noise. The idea is that subjects do not always intentionally and strategically scan, monitor or supervise their environment in search of knowledge, but are influenced by a mixture of rumours, interpretations, impressions and recommendations (Asheim and Coenen 2007: 658, Grabher 2002: 209). Although buzz is often characterised as local buzz, it can be transferred globally via the media or by travelling professionals. Buzz therefore does not necessarily imply face-to-face interactions and physical co-present communication (Asheim and Coenen 2007: 657–8). Rather, it is linked to communities of practice (Wenger 1998) which act face-to-face as well as at a distance, even globally (Gertler 2008, Jones 2008, Roberts 2013: 88).

Implicit Knowledge

The term 'implicit' knowledge is widely used alongside 'informal' knowledge. Like informal knowledge, 'implicit knowledge' is often defined by features describing what it is *not*. Implicit knowledge does not refer to explicitly articulated mental constructs. Implicit knowledge is the tacit or unspoken knowledge carried by persons or groups (Polanyi 1958, Nonaka and Takeuchi 1995), thus representing the hidden part of knowledge, the part which is not explained in manuals and can only be captured with difficulty, if at all. Implicit knowledge is largely subconscious or subliminal.

   Usually it is not necessary to discover and articulate implicit knowledge (Arena, Festré and Lazaric 2012: 3). Often, implicit knowledge is learned through observation and experience, much like learning to ride a bicycle which also needs to be experienced rather than explained (Witt, Broekel and Brenner 2012: 329). Usually, implicit knowledge is unreflected, and most people find it difficult to express it verbally. For a technician for example, it is difficult to describe how to run a machine or how to identify flaws in an engine just by listening to it. New colleagues introduced to these tasks need to develop their own feel for the machine, for example by paying attention to the sound of the machine and learning when it is running smoothly.

   Obviously, workers, companies and the respective regions need more than formalised and explicit knowledge. The above has emphasised the importance of scientific-technical knowledge as well as knowledge 'beyond' scientific-technical knowledge which is informal and implicit. Informal and implicit knowledge is often acquired through experience.

Experience

'Experience' means knowledge formed on the basis of former impressions and perceptions. Experiences share some of the characteristics of informal and implicit knowledge in that they are usually not representative, not easy to replicate and difficult to calculate. Neither are they generated under exact settings in laboratories or follow stringent methodological principles (Knoben and Oerlemans 2012).

   At the same time, experiences are not arbitrary. They are structured by memories, interpretation, intentions and plans, and contribute to the restructuring and reproduction of other mental constructs. Experiences thus are controlled, but not by scientific methods; they are only checked superficially, or not at all, or only in case of problems. People usually avoid examining and verifying their experiential knowledge: Soeffner (2004: 15, 19) for example points out that any regular testing of people's knowledge about their colleagues, family and friends would probably irritate and spell trouble with the objects of such examination.

   Experiences are sound, obvious and evident, with their own particular 'rationality' and effectiveness in specific situations (Foray 2012: 267). They are rules of thumb which help orientation and decision-making in complex and confusing situations. In companies, experiences comprise large parts of the know-how (and know-what, know-who, know-why), and the skills required for problem-solving, including particular expertise and specific capacity.

Groups, communities, organisations and societies all share certain experiences. Partially, these are based on direct individual experience ('primary experiences'), but in larger societal contexts, 'secondary experiences' also play a significant role which are often generated by the media. Experiences of natural disasters or of terror attacks in other parts of world are examples of secondary experiences.

Table 4.1 gives an overview of the characteristic differences between scientific-technical knowledge and experiences.

Table 4.1   Scientific-technical knowledge and experiences

	Scientific-Technical Knowledge	Experiences
Formalisation	Formalised	Informal
Explicitness/Implicitness	Explicit	Implicit
Scope	Specific	Broad scope
Representativeness/ Soundness	Representative, replicable and as objective as possible	Sound and evident; situational, individually adapted, partially replicable
Predictability/Calculability	Calculable	Only calculable in few cases, contextualised, 'soft' rules of thumb
Production	Generated under controlled conditions in laboratories or test centres	Generated in biographical or common history, often occasional, serendipitous, contingent
Institutionalisation	Institutionalised in research centres/departments, generated by professionals	Generated in contexts and situations
Controlled by	Methodological standards	Tested only in exceptions and superficially
Evaluated by	Formal organisations	Informally evaluated by further experiences

In daily work life, scientific-technical knowledge and experiences intermingle. Knowledge of production processes and the use of machinery for example combine scientific-technical engineering knowledge and practical use-related knowledge. Usually this mixture of knowledge is at the heart of day-to-day problem solving in companies and by individual employees. The same applies to R&D. Experiential knowledge is important for the production of scientific-technical knowledge even in globally distributed teams of engineers and technicians (Oshri, van Fenema and

Kotlarsky 2008). Normally, this mélange of scientific-technical knowledge and experiences is quite a useful practice.

Chapter 5
Beyond Knowledge:
The Relevance of Interpretation

The previous sections discussed scientific-technical knowledge and further mental constructs as examples of mental concepts which most would consider 'true'. This chapter now turns to mental concepts 'beyond knowledge'. Rather than parapsychology or UFOs as sometimes intimated by the rainbow press, 'beyond knowledge' refers to interpretive cognitive constructs. 'Beyond knowledge' is concerned with aspects of interpretation, such as assessing, estimating as relevant, or viewing in the light of personal feelings and socially mediated judgements.

Even though there is a difference between knowledge (which is considered true by a group, community, organisation or society) and mental constructs 'beyond knowledge', it must be stressed that both are closely related. Knowledge influences interpretations and vice versa. For example, knowledge of a new investment opportunity goes hand in hand with interpretations that influence the eventual decision. Similarly, the interpretation of a new investment destination as a rising star is linked to particular knowledge about this destination.

The idea of mental settings 'beyond knowledge' is not new. Hayek (1937, 1945) highlighted the role of procedural knowledge formation, use of knowledge and the importance of personal interpretation in explaining the functioning of markets. Generally, the role of mental constructs in shaping patterns of economic and spatial behaviour were intensely debated in the 1950s and 1960s, influencing later psychology-based economic approaches to decision-making (for example Johnson-Laird 1983, Rouse and Morris 1986, see Rizzello and Spada 2012: 144–8).

Boulding for instance stated that human action is based on mental representations. He stressed that every person and organisation is not only guided by their mental images and the subjective knowledge they hold about the world, but also by their temporal context, their personal relationships and emotions, and the factors determining such images, such as past experiences and new messages, all of which influence existing mental images by adding further information and values (Boulding 1956: 3–11, see Patalano 2012: 121). Boulding concludes: 'The image not only makes society, society continually remakes the image' (Boulding 1956: 64).

The notion of 'social construction of reality' (Berger and Luckmann 1966/1991) was therefore already in the air. Simon (1957: 241–73) is another early author who discussed the relationship between knowledge and uncertainty and human rationality and decision making (Rizzello and Spada 2012: 147).

In geography, it was Loewenthal (1961) who brought up experiences and imagination as key elements of a spatial epistemology. Today, 'imaginaries' are used to describe the influence of mental models on regional policy and urban planning. Such cognitive settings play an important role in current human geography (Daniels 2010, Gregory 2007) and spatial planning (Boudreau 2007, Wetzstein and Le Heron 2010, 1905–6) where they describe the normative influence of shared mental concepts. Imagination is also important for the entrepreneurial theory of a firm and can help to understand 'cognitive leadership' (Witt 1998). Recently, apart from images and imaginaries, and the recent rediscovery of 'perception' (Crawshaw 2013), 'visuals' have come into focus as a form of immediate symbolic communication (Meyer et al. 2013, Rose and Tolia-Kelly 2012), for instance in design, accounting, marketing, sales and further company activities (Justesen and Mouritsen 2009).

In organisational studies and economic geography, such concepts are often called 'shared visions'. Shared visions are situational views commonly held by a community or members of an organisation. The sociological concept of 'patterns of interpretation', in contrast, refers to long-ranging structures and widespread patterns of interpreting the world. Although 'shared visions' and 'patterns of interpretation' have in common their shared nature – rather than representing the single and solitary opinion of an individual – they differ with regard to their stability and extent.

Shared Visions

Shared visions have been a topic in knowledge research for some years, especially in organisational studies. Recently the concept has also been taken up by economic geography (Bathelt and Glückler 2011). 'Shared visions' represent the shared views of actors, such as managers (top managers, plant managers, R&D managers and other executives), engineers and technicians, workers and 'regional' actors (members representing different associations which shape the region). Shared visions develop in interaction through a shared process of learning, implying that shared visions arise in response to situations and requirements limited in time and space. This makes shared visions a close relative of immediate strategies, plans and practices held by actors.

In a globalised economy, decisions particularly depend on the shared visions of management with respect to company objectives, the ongoing process of globalisation, and the problems and potentials involved in processes of internationalisation. Such shared visions do not directly depend on the requirements of the value chain; nor can they be seen as a simple reflection of locational circumstances (Fuchs 2005). Managers create a shared vision of strategies and actions by communicating their subjective impressions. For example, shared visions can support the improvement of quality standards in the production process (see Chapter 12).

In organisational studies the discussion of shared visions can be traced back to the 1980s and 1990s when the discipline looked at team capabilities, expanding its research to include commonly shared cognitive models and mindsets (Cannon-Bowers and Salas 2001, Strube et al. 2005). Prahalad and Bettis (1986) for instance study the dominant logic of management as mindsets based on experiences, whilst Cannon-Bowers and Salas (2001) discuss shared cognitive constructs. They distinguish knowledge which is *precisely* related to tasks from knowledge which is *broadly* related to tasks, as well as the knowledge the team members have of each other and the shared beliefs they hold. Denzau and North (1994) refer to shared mental models which are essential for interpreting reality and for anticipating the future. Despite the plethora of terms, 'shared vision' is quite commonly used. Today, mental models are a subject of interdisciplinary discussion, for example in organisational studies (Mohammed and Dumville 2001), learning sciences (van den Bossche et al. 2011) and psychological research about social cognition (Lau, Chiu and Lee 2001).

The main point is that shared visions enable actors to overcome problems by developing adequate strategies. Shared visions relate to situations and usually only exist alongside a problem. Shared visions are thus shared mental constructs related to particular situations and specific problems. They often correspond to communities of practice, with the community sharing explicit and implicit knowledge and solving the problem more or less intentionally (Storper 2008). For example, two colleagues – both technicians – share their experiences with a mechanical problem. One of them knows about errors in the electronic system while the other has observed defects when the machine was running. Together, they are able develop a shared vision of the problem which enables them to repair the machine (Hecker 2012: 426).

Shared visions are dynamic, and new situations and problems can require fresh shared visions. For example, in the process of takeovers of firms in foreign countries, shared visions of the parties involved can change over time, as Alvstam and Ivarsson (2014: 230–34) explain for the ownership transfer of Volvo Cars to China. In the early stages of the acquisition, the company was in a 'grey zone' of partially being foreign and partially domestic. Over time, the shared visions changed and led to a clearer picture of self-identification.

Shared visions are different from patterns of interpretation, although the two are also closely related, as suggested by the star of knowledge and interpretation. Shared visions are a form of short-term interpretation, related to situations and particular problems, and therefore dynamic. Patterns of interpretation in contrast provide insights into long-term, structural interpretation and socio-cultural rules.

Patterns of Interpretation

The German sociologist Ulrich Oevermann (1973/2001a, 2001b) proposes the idea of 'patterns of interpretation' as a concept which reaches beyond the notion of knowledge. In a nutshell, patterns of interpretation are the commonly shared, broadly distributed, normative and stable views and considerations held by a society. Patterns of interpretation are persistent and collective socio-cultural 'interpretive schemes' (Giddens 1993: 113), in other words, the rules which direct how meaning and understanding is generated. At present, the approach is one of the most visible, vibrant and widespread views in qualitative research in German-speaking countries (Reichertz 2004), although the discussion has only partially arrived in the Anglophone discourse to date. One example is Wagner, Lukassen and Mahlendorf (2010) who suggest using patterns of interpretation in the analysis of industrial relationships and applying the concept to management and organisational studies.

Structural Patterns, But No Universal Deep Structures

In his early papers, Oevermann (2001a, 2001b) stresses that patterns of interpretation are 'epochal' in the sense that socio-cultural patterns exist for long periods of time. One example is the protestant ethic (see Weber 1920/1986). At that time, Oevermann was inspired by structuralism and linguistic theories (Reichertz 2004). The notion of wide-ranging, elementary patterns of interpretation is reminiscent of 'deep structures', which act as the regulatory units of language. Deep structures can be understood as socio-cultural patterns of symbols or symbol systems people are born into, with little capacity to reflect on, control or change them (Adolf, Mast and Stehr 2013: 29).

Oevermann (1973/2001a: 8–19) has always stressed that patterns of interpretation are not universal structures, but patterns of understanding which are open to development. This view is the result of an intense debate over the last decades. Since the 1980s, the universality of structures and the methods available for identifying distinct and unambiguous rules of transformation between 'surface' and 'deep structures' has been called into question, causing a 'crisis of representation' (Flick 2009: 19). Still, Oevermann's analysis refers to the underlying structures and general nuclei of cultural and social life, focusing in particular on cultural and social fundamentals and essentials. Oevermann (2001b: 41–51) explicitly sees patterns of interpretation as part of 'World 3' (Popper 1967/1983: 68): although patterns of interpretation are part of individual persons, they are also independent of the becoming and passing away of the single individual. Patterns of interpretation work as 'hidden rules', unspoken 'everyday theories' and 'imprints' or 'scripts', although subjects have some – limited – autonomy in how to interpret, understand and construe reality.

Patterns of interpretation work implicitly as long as they provide an inner logic and consistency, offering appropriate and adequate interpretations of situations, social settings and the self; they especially come to the fore in critical situations (Oevermann 2001b: 36–8, Soeffner 2004).

With its focus on structures, Oevermann's view recalls other theoretical concepts. As a normative setting 'in the mind', the concept is obviously related to the long tradition of understanding institutions as 'transmitted institutional habits of thought' (Veblen 1910, North 1991) and internalised rules of conduct (Dutraive 2012: 106–7). Patterns of interpretation also show some similarities to the idea of 'memes', which are taken to be analogous to genes and represent cognitive units, memory elements and socio-cultural patterns of meaning (Dawkins 1989). The term 'framing' is used in economic science to describe the influence of norms, habits and expectations on decision making; in the social sciences, the same term describes cognitive settings or individual and collective regularities which shape perception and knowledge (Arena, Festré and Lazaric 2012: 7–8).

Case-Specific Knowledge and Socio-Cultural Interpretive Schemes

Oevermann (2001b: 39–45) differentiates between case-specific knowledge on the one hand and patterns of interpretation as socio-cultural interpretive schemes or the rules that generate meaning and understanding on the other. Evidently, there is a difference between the subject matter of knowledge and the process of understanding (Egidi 2012: 204). Similarly, Nooteboom (2012: 339) stresses that cognitive categories give structure to perception, interpretation and evaluation. Apart from content, understanding comprises the cognitive 'absorptive capacity' that enables actors to understand the code and interpret its meaning (Witt, Broekel and Brenner 2012: 370–74). A situation of decision-making about relocating production towards an investment area overseas can serve as an example. Managers have particular *case-related knowledge* with much detailed information about the foreign investment area. At the same time they are directed by *socio-cultural interpretive schemes as the rules that generate meaning and understanding*. Such schemes and rules allow the managers to understand the available information per se. Additionally, managers contextualise the information previously interpreted through further cognitive schemes, for example the 'foreign' country, 'risks' of the investment and 'trust' in the partner overseas. All these deliberations illustrate that knowledge not only comprises content, but also shared structural principles of interpretation and heuristics.

Structure and Practice

Some approaches not only focus on structures, but systematically include practices. Giddens (1993: 113) regards 'interpretive schemes' as implicit 'background knowledge' in cultures and societies. This view distinguishes between such socio-cultural patterns and practices, as implied by the notion of social con-'struction' of reality. Giddens (1984: 1–40) takes up this perspective in his prominent structuration approach which combines structure and agents and which refuses to exclusively focus on either. He suggests a perspective which allows social practices to be studied as a series of acts realised by actors, as well as 'forms' representing structures pertaining to collectives or social communities (Giddens 1993: 110).

From the perspective of practice-oriented approaches, it is odd to ask 'what *is* knowledge' because a subject always *knows* something (Ibert 2007). Talking about knowledge means talking about subjects involved in a process of cognitively appropriating the world. This makes knowing a practice which is always related to learning. Practice such as speech is the 'completion' of a thought, and not simply a consequence of a pre-existing mental concept (Fenton and Langley 2012, Hinchliffe 2000: 576). In this view, knowledge is always expressed in practices through language and other symbols (Martin and Moodysson 2011: 1186). Language and other symbols are not pre-given structures, but can change by communication over time. The research agenda of practice-oriented approaches is to identify change, the unforeseeable and the diverse possibilities which may occur (Helbrecht 2011: 117–19). Such approaches have given insights particularly into the mobility of cognitive constructs, and especially the 'translation' of knowledge while it is travelling. As Helbrecht (2011: 115) points out, 'travelling theories' for knowledge fundamentally differ from a pair of shoes travelling around the globe, because travelling theories substantially change on their way across borders, while shoes and other commodities generally remain the same.

The practice-oriented approaches suggest there is no obvious sharp contrast between patterns and practices (as the 'star of knowledge and interpretation' may be thought to suggest). Rather, patterns and practices are the endpoints of the ray which describes the extension and range of knowledge and interpretation.

Synopsis: Shared Visions and Patterns of Interpretation

Given the interrelatedness of shared visions and patterns of interpretation, it is clear that no hard and fast boundaries can be drawn between these concepts. For example, situational shared visions and wide-ranging patterns of interpretation cannot be distinguished by referring to particular and measurable time spans. Oevermann does not consider the range and stability of patterns of interpretation to be a given, suggesting instead they need to be determined in each case by research. In a critical contribution, Ullrich (1999: 429) accepts that patterns of interpretation refer to the overall culture but notes they can appear on different scales and in different parts of society. Kassner (2003) understands patterns of interpretation as phenomena in a 'social field'. Considering the different range of interpretations, there is also some resemblance to 'small stories', to 'master stories' and 'grand narratives' (Fenton and Langley 2012: 1175). In a similar manner, Ferreira, Vieira and Neira (2013: 58) understand culture as a collective memory of a community or a society with different layers. The inner layers are the hidden, secret and stable patterns, while the outer layers are the visible, observable and changeable patterns. Obviously, then, there is a transition zone between the different types of patterns (or 'layers'). Shared visions add the notion of short-term, situational interpretation, whilst patterns of interpretation provide insights into long-term, structural interpretation. Shared visions are embedded in a socio-cultural context and thus in patterns of interpretation; shared visions depend on the backstage rules which the patterns of interpretations represent.

The case studies will demonstrate the joint influence of shared visions and patterns of interpretation on the thought and action of managers and other actors (see Chapters 11–13). Table 5.1 summarises the relevant characteristics of shared visions and patterns of interpretation.

Table 5.1   Key features of shared visions and patterns of interpretation

	Shared Visions	Patterns of Interpretation
Key academic context	Organisational studies	Sociology
Most relevant way of interpersonal transfer/ learning	By experience	By habituation
Scope	Some actors, for example in an organisation, in a region, in a community of practice	Society, culture
Duration	Particular situation, specific problem-solving	Long time-span
Structure/practice	Rather related to practice, strategy and decision-making	Structural patterns
Normativity	Normative solutions for problems	Socio-cultural rules
Implicitness	Explicit/visible or implicit	Implicit, hidden rules, backstage rules
Interrelationship of shared visions and patterns of interpretation	Embedded and dependent on patterns of interpretation	Framework for shared visions

The main overall benefit of shared visions and patterns of interpretation is their ability to broaden the perspective beyond knowledge. Economic decisions, strategies, activities, and the resulting institutions do not simply arise from knowledge or external economic preconditions. They are particularly a product of interpretation. This comprehensive perspective therefore helps geography to better understand spatial socio-economic phenomena and their evolution (MacKinnon et al. 2009, Martin and Sunley 2006).

Chapter 6
Knowledge and Interpretation in Temporal Dynamics: Learning

In a nutshell, learning is usually considered the acquisition, adoption and appropriation of knowledge, shared visions, patterns of interpretation and other cognitive constructs. Based on the previous deliberations on practices and the rules involved in generating meaning and understanding, learning can be defined as meaningful action.

In this reading, the learning subjects are the members of the organisation, including management, employees, or workers' representatives – in other words, those who have acquired knowledge or other interpretations and have the competence to use it. This kind of learning can be a convoluted process in the 'multiple organisation' of large and multinational companies (Wiesenthal 1995).

In a different understanding, many publications also refer to learning in terms of company organisation. If related to subsidiaries strongly integrated in value chains, this process of learning is termed upgrading. Here, interactions with the multiscalar environment play a role which is understood in this context as actor networks and institutional arrangements. This chapter explains learning as meaningful action in the work process. It first highlights the particularities *of learning members* in multinational companies. Then, the perspective shifts towards organisational learning in *multinational companies*, with particular focus on multinationals' foreign investments and the role of the subsidiaries in multiscalar environment. In the subsequent discussion of the globalisation of knowledge, Chapters 11–13 refer to the learning members of the organisation, such as the learning of management (for example how to internationalise production and R&D) and the learning of employees (for example vocational education and further training). Chapters 11–13 also relate to organisational learning (such as upgrading towards integrated production plants). There is no contradiction between the learning of the members of an organisation and the learning of the organisation since the concept of 'learning' depends on the particular context. Empirical evidence demonstrates that learning members of an organisation and organisational learning often go hand in hand.

Learning as Meaningful Action in the Work Process

In daily life learning serves to acquire, communicate, preserve, use, complement, refresh and refine knowledge. Much like eating and sleeping it is a life-sustaining and inevitable activity (Wenger 1998: 3) which is part of the ongoing social construction of reality (Berger and Luckmann 1966/1991). Thus, learning is meaningful action. Out of the many facets of the term 'meaningful', three denotations are significant here.

Firstly, from a sociological perspective 'meaningful' action implies 'intentional' and 'motivated' action. Meaningful action is thus understood as a contrast and opposition to 'rule-governed behaviour'. While 'behaviour' is the aspect of action which can be observed by others, meaningful action is always meaningful and intentional to the acting person. For example, to wave one's hand is an observable behaviour; to the subject waving their hand this is a meaningful action which might serve to greet somebody or wave farewell (Fischer 2012: 37–8).

Other persons watching the subject wave their hand can understand such action. 'Meaningful' in this second sense of the term refers to similar structures or commonly shared patterns of meaning (Fischer 2012: 37–8). Even though there may be different motivations for a person waving their hand, the person greeted or waved off will probably understand the intention. Other observers will relate to the same shared patterns of meaning and will also interpret what they perceive.

Patterns of meaning can evolve over time, for example in a work team. Creating new or modified meanings which have not previously existed in the team is a process of mutual co-construction. For example, the implicit or explicit rejection of the prevalent understanding of a meaning in a group can lead to the creation of new shared meanings. In work teams, through negotiation by argument and clarification, the team gradually develops convergent meanings and thus commonly shared mental models (van den Bossche et al. 2011: 287–8).

The third denotation of meaningful refers to the satisfaction which the action generates for the acting person. This interpretation of 'meaningful' is closely related to work life. For example, paid work not only leads to the satisfaction of receiving a salary, but also the satisfaction of solving problems, of putting ideas into practice, or of being appreciated. Work tasks requiring comprehensive knowledge, training and learning play an important role in the perception of labour as meaningful.

Job design and the structuring of work are closely related to learning as meaningful action (in the sense of the satisfaction experienced when accomplishing a task and self-confidence). Institutional arrangements can facilitate the creation of meaningful work. Labour law offers a basic framework for this, to which further regulations are added such as collective agreements and company agreements. Such regulations set out the extent of education and training available during working hours, the scope of paid educational leave, or measures linked to the job, such as job rotation, job enrichment, group work, company suggestion systems, and other means of participation and co-determination.

Learning as meaningful action can have emancipatory power, as the proverbial expression that 'knowledge is power' claims. Knowledge has the potential to influence power dynamics (Marston and de Leeuw 2013: vi), offering the opportunity to first become aware of and then potentially overcome hierarchies, entanglements and obfuscation, at least to some degree. Labour history and the history of social movements illustrate the importance of emancipatory knowledge and counter-knowledge in doing so. In his three-volume novel *The Aesthetics of Resistance* (2005), set during the period of fascism, Peter Weiss illustrates how the narrator, a worker, suffers due to his limited opportunities for learning whilst simultaneously needing education and knowledge in order to be able to interpret reality and act in the European antifascist movement.

At this point, access to learning becomes an important issue. Exclusivity has already been discussed with regard to scientific-technical knowledge, but the same applies to learning of employees, as opportunities for education and vocational training are often exclusive. In many societies and communities schooling and professional training is a privilege exclusively enjoyed by men and hardly accessible to women. Children from high-income households and with parents who place great value on education usually have better access to learning. Access to higher education is still a problem for the less educated migrant population in many parts of the world, too.

Hence, access to knowledge is important for workers, workers' representations, as well as social movements. In order to change the prerogative of interpretation, such organisations and their representatives and activists need to be particularly well-informed and intellectually prepared (Goodman, Boykoff and Evered 2008, Novelli and Ferus-Comelo 2010: 50–54). Hosseini (2010: 25) stresses that 'dissident knowledge' forms as part of a continuous process of developing awareness of particular social issues. Access to knowledge is necessary to break the power of definition held by the dominant discourses and imaginaries – and to develop alternative shared visions instead (Sacchetti 2009: 3, Sacchetti and Sugden 2009: 271). This is true in a broader political sense as well as in diverse constellations, perceived discrepancies and (subliminal or open) conflicts. Hence, learning is framed by a wide range of individual interests, social contexts and situational settings, particularly in multinational companies.

Learning in Multinational Companies

In multinational companies, managements' problems are related to the complexity of large organisations where employee qualification and training may be difficult to control (Schmid and Daniel 2011). The 'multiple organisation' (Wiesenthal 1995) of large and multinational companies challenges managers in the implementation of targeted training programmes. In such globalised contexts, 'islands of expertise' play an important role.

Islands of expertise can cause problems for managers due to the ambivalence of multi-directional and multiple learning processes. On the one hand, unplanned processes of learning are necessary in order to keep operations running, allowing for different potentials, strategies and ways of learning in the different parts of the organisation. On the other hand, learning processes which only incompletely follow top managements' objectives and the hierarchical structures of the organisation, can create poorly integrated 'islands of expertise' with a high degree of autonomy (Glückler 2011: 207). This leads to difficulties, for example if knowledge is to be shared with other departments, or if such knowledge is lost when the knowledge holders leave the company. Thus, management has to ensure that the different units pool and merge their knowledge (Calantone, Cavusgil and Zhao 2002: 515–17, Revilla and Rodríguez 2011).

Technological expertise is considered especially prone to creating such islands. Formal hierarchies tend to fail if management simply imposes top down directives as managers are dependent on the cooperation and exchange of know-how with experts. Hierarchical coordination needs the 'benevolent' motivation of the actors (Scharpf 1997: 174–8), turning especially the management of labour relations in R&D into 'the art of compromise' (Segrestin, Lefebvre and Weil 2002: 68). At the same time, the gap between managers and technological experts is not the only one to be bridged. In companies, differences also exist among the various technological experts. Ibert (2010: 199) illustrates such ambiguities by looking at the allocation of knowledge among the different actors involved in designing a sensor system. In this example, the 'scientists' depend on external knowledge and travel a lot, but they mostly tap into external knowledge at a singular location and thus create a concentric spatial pattern. 'Developers' dealing with application procedures also need a lot of external knowledge, but they act in a polycentric environment. Whilst such differences may simply reflect organisational realities, supporting the overall goal of efficiency, such processes may also develop their own dynamics, leading to the further decoupling of islands of expertise within the company and thus weakening the common knowledge base of the organisation (Alnuaimi, Singh and George 2012). Veto-holding actors may also be created (Scharpf 1997: 174–8).

Cooperation with respect to technological expertise is particularly important in transnational companies with globally distributed know-how centres. Shared learning requires critical reflection from different perspectives, bringing in new concepts, opening up new lines of investigation in new fields (preventing 'not invented here' effects) and helping to find systematic principles, separating 'the wood from the trees' (Bessant et al. 2012: 1091). It follows that international companies need shared learning in order to create commonly 'shared visions'. Shared learning requires certain preconditions, such as the willingness and ability to overcome linguistic barriers, tolerance with regard to world views and political differences and a general atmosphere of openness which helps to overcome cultural ascriptions (Karlsson and Johansson 2012: 30–31). Often, learning includes collaboration between individuals over a time span; then, learning means the gradual development of shared mental constructs (van den Bossche et al. 2011: 286). Thus, a transnational company needs to deal with different shared visions and patterns of interpretation, countering superficial categorisation as 'the others' which can hamper cooperation.

Localised Organisational Learning in the Multiscalar Environment

This section shifts the perspective from individual learning in companies towards organisational learning. From the evolutionary perspective of organisational learning, the subsidiary of a multinational can be described as interacting with a multiscalar environment, which itself can be characterised through actor networks and institutional arrangements. It is the dynamic interaction with this environment which can lead to processes of organisational learning and upgrading.

Localised Interaction

Learning organisations in multiscalar actor networks and institutional arrangements are dynamic entities which interact with their environment. This follows the notion of Manning, Sydow and Windeler (2012: 1202–3), who criticise the view of transnational companies as either passively embedded in local networks (the network acting as a refuge) or as exploiters of local conditions (in line with their own strategic goals). In their view, transnational companies should be regarded as actors which actively connect the company to other companies and multiscalar organisations and institutions through context-specific collaborative and competitive practices that become adopted across locations and regions. They emphasise that transnational companies actively select those locations for their international subsidiaries which offer qualified labour, supporting organisations and institutions and peer firms with similar demands, and then actively shape these locations.

This is a useful development of the notion of 'adaptation' as used in evolutionary economics and evolutionary economic geography. In the evolutionary context, 'learning' describes the adaptation of an organisation to a permanently changing environment (Argyris and Schön 1978). A more recent concept addressing how companies interact with the economic, social and political environment and vice versa, is that of co-evolution (Boschma and Martin 2010, Cantwell, Dunning and Lundan 2010, Dosi 2012: 173). The co-evolutionary perspective highlights the close interrelationship between companies and their environment which gives rise to the insight that both – company and environment – will change together gradually.

In some cases though, a particular setting can induce a period of stagnation and lock-in (Hassink 2005) which can end in the 'atrophy of capabilities' (Birkinshaw and Hood 1998: 783). Consequently, 'unlearning' (Ernste 2003) and 'forgetting' (Carmona and Grönlund 1998) seem to be opportunities for the organisation and its environment to overcome such lock-ins.

Actor Networks and Institutions

Depending on the context, actor networks refer to a range of actors including organisations and individual subjects. Thus, an actor can be a multinational company, a particular subsidiary, a global non-governmental organisation, a governmental organisation, a trade union, an employers' association, a chamber of trade and commerce or a business development corporation. An actor can also be an individual person, such as the representative of an organisation or an influential person in the organisation. Often, actor networks are multiscalar in that they comprise the global, the supra-national, national, subnational–regional and local level. The relationships within actor networks, the strength of the ties and the frequency of contacts are subject to a broad debate (Bathelt and Glückler 2011). The empirical chapters will illustrate the role of actor networks in organisational learning and vocational training with regard to upgrading and local development (Chapters 11 and 12).

Actor networks are not accidental; they are shaped by institutional settings. Institutional settings are the norms, regulations, laws, guidelines and formal and informal rules that guide behaviour and action. Institutions can be seen as the 'environment' which frames the elbow room available to actors and actor networks; at the same time institutions are 'embodied' in the actors and permanently reproduced and restructured by daily practices (Veblen 1910: 68, North 1991, Denzau and North 1994: 4, Djelic and Quack 2008: 300). Institutional views particularly seek to explain how regulation works, in other words, how institutional arrangements frame actor networks and vice versa (Fuchs 2012). In the context discussed here, the key role of institutional settings is that they frame actor networks and particularly regulate the conditions for learning and upgrading.

Upgrading

In the last decade, upgrading has become a key concept in the discussion of organisational learning, in particular in political economy, industrial sociology and economic geography. Multinational companies, as well as global production networks, show diverse distributions of power, mandates, and competencies and are thus not adequately described by simple patterns of centre and periphery. Various relationships need to be accounted for, as well as the continuous dynamic of upgrading and downgrading (Gereffi 1999, Schmitz 2004a). The key focus of the upgrading debate is on the ways in which subsidiaries, as at least formerly 'peripheral' plants of transnational companies, are able to 'move up' by capturing and augmenting value and/or new competencies. Plant upgrading is characterised by product innovation, process innovation and functional upgrading, such as receiving new departments and tasks, or by moving into higher value-added activities such as design, branding, marketing and retailing (Tokatli 2013: 1008). Additionally, there is inter-chain upgrading into more efficient production chains (Schmitz 2004a: 7–8, Werner 2012: 404). Thus, upgrading implies a gain of knowledge as well as additional responsibility and power (Pietrobelli and Saliola 2008). As such, upgrading refers to 'competencies', comprising the knowledge, the responsibility and the capability to act and also including sufficient elbow room for further learning within the organisational unit.

Upgrading crucially depends on coordination within the multinational company and the global value chain. In such global production networks, different kinds of coordination exist which extend beyond market prices and hierarchies. In 'modular value chains' for example, the supplier follows the exact technological specifications of the client, whilst in 'captive value chains', a strong lead firm captures the supplier via contracts (such as the Japanese-style production networks). Last not least, 'relational value chains' play a role, which are characterised by mutual interdependence, and which offer better conditions for subsidiary upgrading than the other variants of chain governance (Gereffi, Humphrey and Sturgeon 2005). Relational value chains have attracted a lot of interest in the academic debate, sometimes as part of the discussion on 'trust' in the economy and sometimes in the context of participation.

External governance by actor networks and institutional arrangements also has some influence. Thus, upgrading is also the result of transnational companies and their subsidiaries interacting with multiscalar organisations, regulations, movements and unions. Formal as well as informal institutional arrangements play a role (Schiller 2013). Governance is orchestrated by diverse actors and institutions that operate on different spatial scales (Phelps and Wood 2006: 494). Currently, a tendency is noted that collective bargaining agreements, international framework agreements and other international institutions frame the working conditions in subsidiaries of transnational corporations even in peripheral locations (see Chapter 12). In the discussion on upgrading and global governance, the underlying normative idea is that global governance is a step towards cosmopolitanism and the 'world society' (Messner 2002: 60).

Recent contributions on global governance highlight the importance of labour relations (Coe, Dicken and Hess 2008: 284–5, Henderson et al. 2002: 447, Jessop and Sum 2006). As an example, Cumbers, Nativel and Routledge (2008) focus on labour agency and union positionality in the context of the global production systems, whereas Selwyn (2012a, 2012b) emphasises employment including the political representation of labour; self-awareness and self-perception of workers play an important role in his approach. Werner (2012: 405) focuses on structures and narratives which produce and reproduce differences between men and women, skilled and unskilled labour in the context of upgrading. This perspective on upgrading is taken up in Chapter 12, which specifically focuses on the role of human labour in the 'peripheries' of the world system.

Chapter 7
Knowledge and Interpretation in Place and Space

The notions of place and space are used here in the sense that actors, actor networks and institutional arrangements are located in particular places and have a particular geographical reach. Such a perspective is rooted in an understanding of space – like time – as an irreplaceable dimension. Immanuel Kant assumed that space and time are given categories of human recognition and perception (Kant 1787/2011: 81–7). Although nowadays geographers rarely refer to pre-given categories, preferring to refer to the social construction of spatiality (Jöns, Livingstone and Meusberger 2010), space is not an object or convertible feature or attribute of things. We can, theoretically, imagine a world without mobile phones or even without the colour red, but we cannot image a world without space (or time) as we always act in space and time.

In the context of 'worldwide knowledge' however, a closer look is needed at how spatiality is socially constructed, and how space and place are discussed with regard to knowledge. There are diverse ascriptions and metaphors for characterising the spatiality of the globalised world. With regard to globalisation, spatiality is often considered 'relational' (Bathelt and Glückler 2011), especially in the light of the 'network society' (Castells 2000). Such relational approaches help to develop a notion of localities beyond concepts of container territories (Jones and Woods 2012). A recent metaphor for space is the rhizome, stressing the strong interdependencies of different locations in global space (Sheppard 2002). Such views are pushed by the insight of the dynamics created by electronic communication networks, and the tremendous growth in mobility enabled by transport infrastructure and energy consumption, resulting in 'time-space compression' (Harvey 1990).

With regard to worldwide knowledge, global firms, local labour and the region, two kinds of spatiality seem particularly important: the interpretation of space as loci of knowledge production and the interpretation of space as dependency and interdependency in knowledge-production.

The Interpretation of Space as Loci of Knowledge Production

In economic geography and regional sciences, the topic of knowledge in place has been debated since the 1980s with particular focus on innovation (Harrison 2013: 56). 'Arguing with regions' relates to a particular interpretation of place, which understands regions as potential high-performers in the worldwide competition (Agnew 2013: 6–7). Contributions refer to universities, research centres, education and training systems, as well as supporting services as a key to the global economic success of firms and regions. The term 'triple helix' has been coined to describe the interaction of government, academia and industry. In the scientific debate important milestones were the idea of national innovation systems (Freeman 1995, Lundvall 1992, Lorenz and Lundvall 2006, Nelson 1993) and the notion of regional innovation systems (Cooke, Heidenreich and Braczyk 2004) and territorial innovation systems (Benneworth and Rutten 2013: 187–9).

At the same time, there was also a broad theoretical and political debate on learning regions, which were described as regions open to change and with coping mechanisms that prevent path-dependencies and lock-ins. Still, such concepts always remained rather vague and could not overcome inherent limitations, neglecting for instance the extra-regional, inter-regional relationships (Sunley et al. 2008: 695). Still, their very fuzziness contributed to the success of these concepts in the political sphere (Hassink and Klaerding 2012, Rutten and Boekema 2012: 982). Thus, today there is a general sense that place-based as well as inter-spatial networks play an important role in knowledge-production, innovation and – a recent topic – creativity. The co-location of 'related capabilities' encompasses different spatial distances and scales (Florida, Mellander and Stolarick 2012: 184).

In the context of such ideas, the view has shifted from knowledge as scientific-technical knowledge towards recognition of other cognitive constructs. The academic debate on places of knowledge production and learning regions therefore parallels the line of argumentation presented above. After the early discussion surrounding the creative and innovative 'milieu' (Moulaert and Nussbaumer 2005: 47) and 'untraded interdependencies' (Storper 1995), a prominent view has recently emerged which describes global cities, metropolitan areas and particular locations within such towns as having something 'in the air' (Belussi and Sedita 2012: 166). The 'regional social capital' (Malecki 2012) and the context-dependency of knowledge bases (Martin 2012a) are considered relevant drivers of innovation.

It is well known that cities are a focus of innovation. Törnqvist (2011: 26–39) describes ancient Athens as an early example of a 'creative city', just like Florence during the Renaissance and Vienna and Berlin in the nineteenth and early twentieth century. The salient point is that these former and current world cities are considered unique and exceptional locations; as such they are irreplaceable and not arbitrary. Today, further dynamics are added by positive feedback loops: regional marketing and planning consciously contribute to the semantic production of such places as creative cities (Boudreau 2007, Faulconbridge and McNeill 2010, Wetzstein and Le Heron 2010). Thus, imaginations hold 'built' and 'constructed' facticity (Harvey 1990).

Important features of creative cities are their 'global talent pool' of highly qualified professionals representing diverse nationalities (Beaverstock and Hall 2012), especially in New York, London, Paris, Montreal and Los Angeles. 'New Argonauts' – referring to the myth of the Greek ship 'Argo' – are thought to build important technological capabilities in distant regions. Often, Argonauts are skilled migrants who are part of Diaspora networks and bring their home countries' expertise to specific industries, thus producing new local–global knowledge networks (Saxenian 2012: 28–30). In such places, there is a multicultural accumulation of educated persons with occupations in science, teaching, engineering, and 'bohemians' who are working in the arts, entertainment, film, multimedia, design and fashion, together with other representatives of the 'creative class' (Florida 2007). There seems to be some interdependence between artistic and cultural activities on the one side, and technology and innovation on the other. Such places become attractive because of the presence of highly educated, skilled people, as well as a particular tolerance and openness to the diversity of ethnic groups, beliefs, values and lifestyles. Three 'Ts' are regarded as important for regional development: talent, technology and tolerance (Florida, Mellander and Stolarick 2008: 616, 645).

However, it is hard to measure the direct effect of creative people on regional economic performance (Marrocu and Paci 2012: 397). Education and inventiveness are considered to contribute generally to an inspiring regional environment, thus acting as a signpost advertising an attractive location (Spencer 2011: 60). The same is considered true for the built environment, such as trendy restaurants, stylish pubs, fancy coffee houses and art galleries in the revived cities. Obviously they contribute to improving the image of places, but their effect on economic development is difficult to quantify (Sunley et al. 2008: 695).

One can conclude that on the one hand, innovation and creativity need a specific location, such as a hip district. Such places are ascribed surplus meaning. The world does not exist 'on the head of a pin' (Massey 1984: 51); and specific localities, with all their complexity, physiognomy, activity, spirit and contingency, are significant for innovation. Such localities signify more than only a 'there'. These places combine diverse, partially causal, partially contingent, and partially random phenomena. Place in this context is a 'veto' against ubiquity and sameness (Schlögel 2003: 10).

On the other hand it is possible to create innovation and creativity 'somewhere'. Professionals often meet at replaceable and impersonal 'off spaces' or 'non-places'. Such places are interchangeable, anonymous and neutral. Managers meet in faceless conference centres in skyscrapers, airports or train stations. Experts use their limited working time travelling together in trains, cars or planes, or quickly meet for lunch in fast food restaurants. Such places are only localities for meetings without an assigned identity or peculiarity. Sometimes actors do not even meet face-to-face. Creative 'buzz' can also occur in space created by electronic or print media (Martin and Moodysson 2011: 1187, Moodysson 2011).

Here the focus is on regions as locations of multinational subsidiaries, where the subsidiaries are integrated in actor networks and institutional arrangements. Such interaction and institutional embeddedness is 'localised', meaning it takes place in the investment region. At the same time, the actor networks and institutions are not exclusive local, implying the subsidiaries are engaged with 'multiscalar' actor networks and institutions.

The empirical Chapters 11–13 will illustrate a further point: some of the new hotspots of global R&D and production are not the creative cities described above and lack their surplus cultural meaning. Instead, they are nodes of global engineering and production networks. Although they are often considered faceless, for example the *maquiladora* cities in Northern Mexico, they are important and specific places for the production, transformation and recreation of knowledge. They are special places, not interchangeable, chosen by transnational companies as locations for their international subsidiaries, because they offer peer firms with similar demands, qualified labour, supporting actor networks and institutions (Manning, Sydow and Windeler 2012: 1202–3).

The Interpretation of Space as Dependency and Interdependency in Knowledge Production

In parallel to the debate on space as a locus of knowledge production, there is another vibrant discussion on knowledge in distant space. This strand of literature, which primarily refers to the context of production and labour in the world economy, has different roots to the above. The discussion was triggered by the increasing growth of multinational companies in the 1960s and 1970s, which generated a division of labour and knowledge. Steering competencies and increasingly qualified labour in the production plants of the core economies came to exist alongside low knowledge requirements and little elbow room for decision-making in the peripheral subsidiaries. While competencies were strongly centralised in the core economies at the top of the company hierarchy, work tasks in the peripheries were extremely specialised and repetitive. Blue-collar workers on the assembly line only needed limited and short training on the job. Such standardisation made the subsidiaries vulnerable as it was easy to shift production further to another country. The discussion of such dependencies, which Lipietz (1986, 1987) later described as 'peripheral Fordism', stimulated further academic and political debate. Multinational companies were assumed to exploit the labour force, raw materials and environment of the 'Third World', often also exerting influence on governments in the Global South. In Latin America the food production sector offered outrageous examples and initiated a debate on dependencies from a political economy perspective.

Despite the simple dualism of core and periphery inherent in this concept, it was long assumed to be adequate for describing global production. Obviously, it remains a suitable model for describing some cases of dependency, for example parts of the garment industry or food production. At the same time, the world is not simply divided into the rich North and the poor South. Elites play a role in the North as well as in the South (Mills 1956/2000). Besides, in Africa, Latin America or Asia, foreign direct investments do not necessarily lead to worse job conditions than those offered by domestic production. On the contrary, there are several cases where workers prefer a job in the subsidiaries of global companies due to the income levels this provides, safety at work, and opportunities for vocational training and education.

An additional aspect is that power relations in the world economy have changed in the last decades. Headquarters are no longer only situated in the traditional core economies, and new centres of power are emerging in China, India, Brazil, and Russia. Such dynamics undermine the simple model of core and periphery in the world system. With regard to markets, Sturgeon and Florida (2000: 12–14, 2004) distinguish different types of regions. LEMA (large existing market areas) include the USA, Japan and Western Europe. Next there are PLEMA, the peripheries of large existing market areas, which include countries located close to large market areas (for example, Mexico and Central Europe). In the PLEMA, manufacturing industries make active use of the lower production costs and the proximity to the markets to produce for the LEMA. Additionally, there are the BEM, the big emerging markets, such as China, India and Brazil, with their own large and expanding markets.

Although such concepts are useful, worldwide relations are complex and inconsistent and do not simply span 'container spaces' such as LEMA, PLEMA and BEM. Different production systems are globally distributed, strongly interdependent and permanently changing, creating multiscalar interdependences (Capello and Dentinho 2012: 4–5). This idea also dispels the myth that the global level shapes the local level. Rather, regionally distributed industrial systems continuously influence and reshape the production networks on different scales, in various spatial relationships and in the diverse places (Lee and Saxenian 2008: 174, Bathelt and Li 2014: 55). At the same time, transnational relations create local places (Manning, Sydow and Windeler 2012: 1202–3).

At this point, we have come full circle, overcoming the interpretation of space as a container. We also overcome the interpretation of space as simple dependency between the North and South. Instead, there are internationally interrelated production systems with multinational companies and multiscalar actor networks which create new places through interaction. Such new places and new global loci of knowledge production are both the drivers and results of such globalisation dynamics.

Chapter 8
Interim Conclusion

Up to this point, a conceptual framework has been developed of knowledge in the global economy. It began with the notions of economic knowledge as a commodity, resource and capital. In this context, scientific-technical knowledge and particularly R&D is considered a key asset as well as impetus for the knowledge-based society. Using the star of knowledge and interpretation as a conceptual basis, this view was then broadened and specified, illustrating that further kinds of knowledge are relevant for adequately capturing 'worldwide knowledge'. Knowledge was explained as a social construct, opening up a framework which was then used for discussing informal and implicit knowledge including experiences. This allowed a perspective beyond scientific-technical knowledge and even 'beyond knowledge' itself. At this point, the concepts of shared visions and patterns of interpretation were suggested as productive theoretical instruments. Although closely related, it became obvious that a distinction needs to be drawn between case-specific knowledge and shared visions as situational, problem-related mental constructs allied to strategy and decision-making, and patterns of interpretation as socio-cultural interpretive schemes or rules generating meaning and understanding.

Learning is meaningful action for employees in the context of their daily work. In multinational companies, management faces particular challenges in terms of organising internal learning processes. Problems can occur of 'islands of expertise'. The view then shifted from individual to organisational learning. Organisational learning was described as the interaction of an organisation (such as a subsidiary of a multinational company) with the multiscalar environment, with the latter understood as localised interactions, actor networks and institutional settings in the particular region. Such interaction can result in plant upgrading.

Finally, the conceptual deliberations discussed spatiality. With regard to economic knowledge, some places are said to have surplus meaning in the sense of particularly 'creative' or 'innovative' places. Space also implies distant interrelationships such as the space of dependency and interdependency.

The next chapter addresses methodological issues before shifting to empirical insights.

PART III
Globalisation of Knowledge in R&D and Production – Empirical Insights

Chapter 9

Studies on the Globalisation of Knowledge: Methodological Introduction

Chapters 10–12 address the internationalisation of knowledge in R&D and production with particular focus on the period since the late 1990s. Although the general debate surrounding the globalisation of manufacturing can be traced back much further, recent years have seen a change in its relevance. The globalisation of knowledge has begun to matter – in the media, to political parties, employers' associations and trade unions. Global knowledge – be it in engineering or on the shop floor – is increasingly important to companies and interest groups. The following chapters particularly use insights from the author's own case studies (Fuchs 2001, 2003a, 2003b, 2005), sometimes conducted together with co-authors (Fuchs and Winter 2008, Fuchs and Scharmanski 2009, Fuchs and Meyer 2010, Fuchs and Kempermann 2012). Further literature is used to highlight recent phenomena, patterns and trends.

Despite their varied research designs, the case studies have in common that they are all framed by explorative and qualitative methodology. This is a consequence of the idea of 'social construction of reality' (Berger and Luckmann 1966/1991) and the insight that research is framed by particular socio-cultural contexts. Research is therefore subject to interpretation (Dühr and Müller 2012: 424) and represents a continuous process of constructing reality (Flick 2009: 19, 63). The choice of an explorative and qualitative approach is also owed to the fact that research on 'worldwide knowledge' is a sensitive issue. Qualitative methods are needed to discover the unknown dimensions of knowledge and interpretation. New empirical trends in the globalisation of knowledge are hardly foreseeable ex ante and thus need to be revealed through an explorative approach.

As the studies were conducted between 2000 and 2013 in different research contexts, there is no homogenous methodological approach, although expert interviews and content analysis are usually the methods of choice. In order to discover patterns of interpretation, the methodology was also guided by Oevermann's (2001c) approach and other forms of 'structural hermeneutics' (Alexander and Smith 2006: 143, Reed 2003: 106).

The following case studies are mostly based on expert interviews with top managers, human relation managers, purchase managers, plant managers, representatives of chambers of commerce, trade associations and other regional actors. Trade union and works council members were interviewed in countries in which they were relevant and where they influenced the processes of learning. Most case studies are from metal-working industries, particularly the automotive,

Box 9.1   Oevermann's methodological approach to objective (structural) hermeneutics

At its heart, Oevermann's method (2001c: 39) is the interpretation of text and wording from within. Analysing written and spoken texts and other communicative material, researchers seek to become aware of the implicit essentials, 'cooling down' their research matter as far as possible. The procedure is a specific method of text analysis which attempts to separate the (largely hidden) rules and structures from case-related knowledge. Patterns of interpretation are identified 'sequentially', first by reading the document exhaustively and thoroughly, then selecting phrases which express relevant structures and then playing through each selected phrase with regard to other versions which might occur. The differences between the varieties help to understand the structures underlying the specific expressions. Throughout this process the researcher is urged to be economical in his work and not fall prey to vivid imagination (Oevermann 2001c: 40).

Although the precise methodology is a clear advantage, it also makes Oevermann's notions vulnerable, not only on account of the sophisticated, time-consuming procedure. One point of criticism is that recent topics, such as issues of globalisation, are rarely 'cold' and that researchers at least partially share the same world as the subjects analysed. The fact that researchers dealing with socio-spatial problems are part of contemporary society and the current academic debate cannot simply be blanked out. Oevermann (2001c: 40) himself admits there is no 'tabula rasa'. His argument is simply that researchers should 'cool down' research matters as far as possible, using criteria such as relevance, explanatory power, evidence, adequacy and intersubjective proof.

Still, because of these difficulties not all scholars follow Oevermann's 'art' of interpretation. Different methods have been set out, combining the analysis of patterns of interpretation with other techniques of analysis (Reichertz 2004) or broadening the approach by combining patterns of interpretation with content analysis and discourse analysis (Keller 2011, Ullrich 1999).

mechanical engineering and electronics industries, including the large lead firms as well as further parts of the supply chain. The selected industries are relevant sectors in the regions examined and are strongly involved in global production networks. In these industries, the duration and dimension of internationalisation are particularly pronounced. The case studies refer to Mexico and Poland, introducing an empirical bias towards 'peripheries of large existing market areas' in America and Europe, as well as Germany as a core economy. Further academic literature is used to overcome the sectoral and regional bias at least to some degree.

It is impossible to give a complete overview of the globalisation of knowledge. The following chapters provide a selective and subjective choice of case studies with particular focus on local labour and the region. The case studies do not claim to be all-embracing or even representative; carried out by the author over a number of years they were not designed as such. They should be understood as illustrative examples, embedded in an outline of recent trends as documented in the current academic literature and other relevant sources. Rather than offer a completed jigsaw, the aim is to make a substantiated contribution to an ongoing discussion.

Chapter 10
Empirical Evidence for the 'Star of Knowledge and Interpretation'

Discussing the globalisation of knowledge in R&D and production in the following chapters, reference is once again made to the categories of the 'star of knowledge and interpretation' (Figure 10.1). The axes of the star organise and systematically broaden the perspective on worldwide knowledge. Over the last decades many multinational companies have learnt how to globalise knowledge. Trends of gradual learning in subsidiaries are particularly pronounced in the metalworking sector, both with respect to R&D (Chapter 11) and the production process on the shop floor (Chapter 12).

Chapter 11 discusses the globalisation of scientific-technical, R&D-related knowledge. Using case studies, the chapter illustrates some indications for increasing sophisticated tasks in peripheries of the world economy. Currently, the internationalisation of R&D generates mixed patterns of 'D without R' and of 'R and D'. Chapter 12 illustrates the globalisation of production-related knowledge on the shop floor and plant upgrading; some international plants have become integrated production sites. However, such processes do not simply translate into general macro-trends of knowledge internationalisation or substantial effects for the Global South.

Chapter 13 discusses the potential impacts of these dynamics on the North, although the implications for workers and engineers in the core economies are still opaque. In principle, the internationalisation of knowledge could cause a race to the bottom, mirroring the relocation of low-skilled production from the North to the South and the subsequent deindustrialisation in some regions of the North. However, there is also some evidence that further industrial transition might result, leading to new arrangements worldwide (Fromhold-Eisebith and Fuchs 2012).

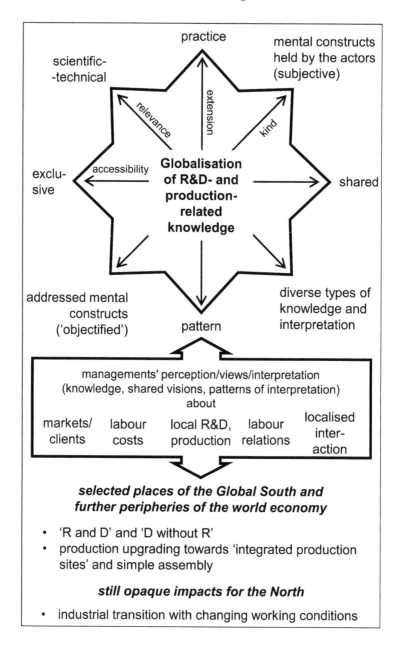

Figure 10.1 The 'star of knowledge and interpretation' applied to the globalisation of knowledge

Chapter 11

Globalisation of Scientific-Technical, R&D-Related Knowledge

In the twentieth century, R&D was considered one of the least movable parts of multinational companies. Recently, however, globalisation has taken hold even in apparently persistent and immovable R&D departments. Centrifugal forces have intensified since the 2000s (OECD 2008: 2).

In line with the broad range of R&D-related activities, the internationalisation of R&D relates to a wide array of tasks. It extends from minor product adaptation and process innovation to radical product innovation or comprehensive systemic process innovation (NIST 2010). The internationalisation of R&D can therefore extend far beyond the implementation of simple design tasks in the peripheries of the world economy. At some multinational sites in the Global South, and in particular places, internationalised activities comprise complex R&D and co-development on a par with activities in the core economies.

The decision to internationalise R&D depends on how management perceives access to foreign sales markets or specific client requirements abroad. Additionally, worldwide labour markets compete with the domestic labour market in a 'global race for talent' or even 'war for talent' (McKinsey 2012a: 14); the 'battle for brainpower' (*The Economist* 2006) is clearly on. Both the product and the labour markets are therefore important in decision-making, with both creating notable dynamics accelerating the internationalisation of R&D. As the case of Huf Hülsbeck and Fürst will illustrate, the increasingly global product markets and the internationalisation of labour markets for engineers and technicians speed up the internationalisation of R&D. Additionally, the particularities of the product development process and the principle of local problem solving means knowledge is increasingly located where it is needed in the world. Last not least, interaction in space, that is a company's cooperation and collaboration with actors and institutions at the foreign destination, also enhances the internationalisation of R&D. In consequence, there are numerous examples of change from 'D *without* R' to co-design and 'R *and* D' in the peripheries of the world economy. Peripheries benefitting from such trends are mainly BEM (big emerging markets) and PLEMA (peripheries of large existing market areas). One of the most notable examples for such internationalisation of R&D is the case of Delphi in Ciudad Juárez in Mexico. Delphi is a multinational company with integrated R&D centres worldwide; the site portrayed here is the technology centre in Northern Mexico. The final sections of this chapter place the dynamics described into the context of more general trends in R&D internationalisation and discuss whether these trends suggest a general change worldwide.

Global Clients and Global Talents

Worldwide markets, clients and labour costs are highly relevant for managements' perception of the global economic world. Product market dynamics, as well as those of the labour market for engineers and technicians, accelerate the globalisation of R&D, particularly when both processes are interlinked (Gassmann and von Zedtwitz 1999: 231). The peripheries of large existing market areas, such as Mexico and Central and Eastern Europe, and particularly the big emerging markets of China, India and Brazil, have high expectations for growth in the aftermath of the global crisis in 2008. They are highly attractive areas offering new product markets, access to lower labour costs and thus new prospects for R&D (McKinsey 2011b).

In the case of *global clients* for R&D, companies tend to internationalise their R&D to the places where the clients are located. Decentralisation of knowledge is required by the governance patterns inherent in different value chains. In 'captive value chains', where a strong lead firm 'captures' the supplier via contracts, foreign clients are able to call for localised knowledge. For example, if an Indian or Chinese automotive brand name producer requires supplier R&D at the location abroad (for example from the USA or Europe), the supplier must establish the requested engineering activities there. In 'relational value chains' with mutual interdependence, clients also frequently ask for local level knowledge. Even in 'modular value chains', where the supplying company follows the specification of the client, some degree of knowledge internationalisation can be necessary. Last not least, anonymous markets with 'arm's length market relations' can also stimulate foreign R&D (Gereffi, Humphrey and Sturgeon 2005). Overall, if customers urgently require R&D abroad, suppliers usually have to internationalise the necessary knowledge.

Such imperative demands spell consequences for the internationalisation of R&D-related knowledge and innovation. In a survey of more than 1,100 executives of multinationals, McKinsey (2011b) finds that 16 per cent of the respondents develop entirely new products for customers in emerging market locations. Even though this share seems small at first glance, it implies that one of six multinational companies develops entirely new products abroad because of foreign market requirements. Still, the majority of companies (40 per cent of the interviewees) do not go as far as innovating entire products. They merely focus on fitting into different regional markets worldwide by pursuing local engineering.

Apart from clients and product markets, markets for *global talent* also play an important role in R&D. Sourcing engineering in low-cost labour markets is a relatively new trend, as R&D has long been centralised at the parent company's headquarters and still continues to be located largely in the core economies. Intellectual property rights are assumed to be safe there. Additionally, the core economies offer a high degree of transparency due to standardised university education. An additional advantage in many core economies is the stability of labour relations. In this manner, implicit and informal knowledge are retained in the company. The implicit knowledge retained by R&D in the advanced core economies represents a substantial obstacle to the globalisation of technological knowledge, as already stated more than two decades ago by Patel and Pavitt (1991: 17–18).

However, today's managers increasingly subscribe to the shared vision that internationalisation of R&D can help to save labour costs. An increasing number of companies therefore take part in the 'global race for talent'. Core-peripheral patterns are increasingly moving towards internationalised R&D, with consciously shared knowledge and lower-paid skilled labour in the overseas markets. 'Access to lower-cost talent globally' (Manning, Sydow and Windeler 2012: 1201) seems to be a general trend in various industrial sectors.

Today, there are mixed patterns of centralised and globalised R&D. The variety of organisational solutions for central and internationally distributed R&D shows that different shared visions can coexist. There is much elbow room for decision-making, and in many companies, there is no clear pressure to globalise R&D in a particular way. There is a wide margin for visions and decision-making, including the risk for management to succeed or fail.

Interestingly, the same set of conditions can lead to different shared visions of R&D internationalisation, both of which can be successful. One example is of two design companies which serve similar market segments and act in similar surroundings in the German automotive sector. One company globalised R&D projects worldwide whilst the other kept R&D centralised, introducing only sales offices abroad (Fuchs 2005). Such broad elbow room once more emphasises that it is not enough to simply analyse 'location factors' as 'determinants' of company behaviour. Actors have the freedom to interpret a situation and to choose a particular route out of the various opportunities presenting themselves.

The following case study illustrates a strategy which kept R&D centralised in the home region until the early 2000s, and then changed towards internationalisation of R&D. Originally carried out in 2002 (Fuchs 2005), the case study was updated in 2013. Like many German metalworking industries, and especially automobile component suppliers, the company concerned had emphasised skilled work for high-quality production in the parent company until the early 2000s. A mere decade later, the strategy had changed in favour of internationalising R&D into new product and labour markets with skilled engineers, especially in the emerging economies. Labour markets thus expanded from the home base in Germany to locations of international R&D. The example illustrates the move from internationalisation of production towards the internationalisation of R&D, with continuous emphasis on qualified local labour. The case study also illustrates a modification in the shared vision of headquarters: in the early 2000s, company management had a shared vision to keep R&D centralised, particularly because of the experienced labour force. A decade later, management favours R&D internationalisation, taking account of the demands of the global clients, a globally skilled workforce and global labour costs.

Huf Hülsbeck and Fürst

Founded in the town of Velbert (North Rhine-Westphalia, Germany) in 1908, Huf Hülsbeck and Fürst develop and produce locking systems for the international automotive industry. The region has early metal-working clusters and a tradition of locksmithing skills, and Velbert has been an important European location for the production of keys and locks since the early eighteenth century. Parts for the automotive industry have been made there since 1920. The company was the first innovator to develop components for the new car access and immobilisation systems 'Passive Entry' and 'Keyless Go' (Huf Hülsbeck and Fürst 2012). Today, the company has more than 6,400 employees worldwide, 1,800 of which are based in Germany. About 400 employees work in R&D worldwide, half of them in Germany. Current clients include German, US, Korean, Chinese and Indian automobile companies.

Initial research in the early 2000s highlighted two reasons for keeping R&D centralised. One was the skilled work required for high-quality production, the other the highly centralised R&D of the German car manufacturers (Fuchs 2005). Large German car manufacturers such as Volkswagen, Daimler Mercedes and BMW have most of their R&D in Germany, and they expected close cooperation between their own R&D and that of their suppliers. The motive of proximity to the client thus led to the shared vision in the supply firm – Huf Hülsbeck and Fürst – that skills and responsibilities should remain centralised. Face-to-face contact and meetings usually took place at the car manufacturer's offices, for example in Wolfsburg, Stuttgart or Munich, and the Director of R&D and some engineers visited their clients' R&D departments on a regular basis.

However, already at that time some R&D skills instrumental in improving the manufacturing process were beginning to internationalise. Parts of maintenance, repair functions and tool design for the production processes were shifted to the foreign plants. This was helped by the earlier internationalisation of production plants from the 1980s onwards, a full two decades before the internationalisation of R&D, which provided management with useful experience. Already then, management had chosen international locations with skilled labour and workers experienced in metalworking industries. Today, management follows the same route, investing in regions where workers and engineers have high levels of competency, particularly in electronics.

By the early 2010s the company had changed its strategy towards well-developed internationalisation in R&D. Apart from still growing R&D centres in Velbert and Günding, Germany, new R&D departments were established in the USA, China, Romania, India, Brazil and Korea (see Map 11.1). The new internationalisation strategy is required by the clients, which comprise the brand name producers of the core economies as well as the domestic brand name producers of the emerging economies.

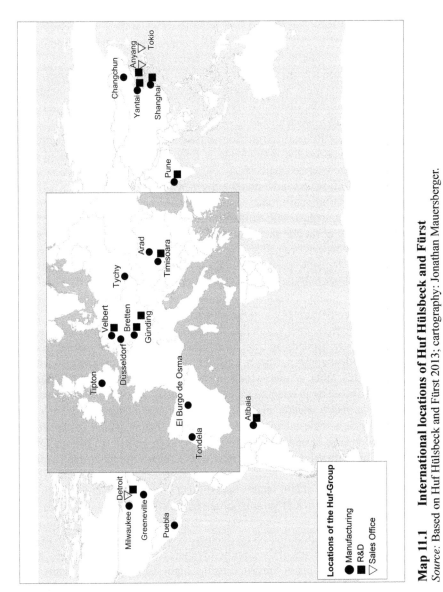

Map 11.1   International locations of Huf Hülsbeck and Fürst

Source: Based on Huf Hülsbeck and Fürst 2013; cartography: Jonathan Mauersberger.

In the emerging economies, domestic car manufacturers such as the Indian company Tata and Chinese Geely require locally designed products. They prefer face-to-face contacts, but above all demand affordable supply parts. With skilled employees for engineering activities still cheaper in the emerging economies, Huf Hülsbeck and Fürst has strengthened its R&D departments there. Responding to client demands for affordable products ('frugal engineering'), they are responsible for the engineering of simplified products and processes.

The resulting products are specially designed for local car manufacturers overseas. As a result, there is no competition with engineering departments in Germany. Due to this clear division of tasks, the company is able to share parts of its R&D-related knowledge with foreign engineering departments without losing innovative knowledge to local competitors abroad. On the contrary, management follows the shared vision that knowledge is only valuable if it is applied and therefore needs to be internationalised.

The case study illustrates three noteworthy aspects.

First, R&D internationalisation is a learning process which is based on managements' prior experience of the importance of skilled labour. While R&D-related labour costs are an important part of the current shared vision of management, other aspects are just as important, such as the availability of skilled labour to ensure high quality production. This is not only the case at Huf Hülsbeck and Fürst, as no automotive supplier dares to ignore skilled labour as a prerequisite for reliable product quality. Thus a combination of low cost and high quality seems to be the most attractive strategy to management (Fuchs 2003b, 2005).

Second, the current shared vision that 'knowledge is only valuable if it is used and applied' strongly depends on trust as an underlying pattern of interpretation. Trust is the confidence in a partner's reliability and integrity, implying positive expectations of the other actors (which in this case are embedded in different international contexts) as an opening for accepting one's own vulnerability. Actors cannot know the exact situational conditions and consequences of all possible actions, so trust is a socio-cultural pattern of interpretation which is applied in cases of shared visions of opaque situations. At the same time, trust is not applied automatically as there might be other cases where the actors doubt and mistrust each other (Li 2005: 77–80).

Third, the case study of Huf Hülsberg and Fürst points to differences between high-road and low-road R&D internationalisation. Globalising R&D does not necessarily imply the same top standards for all market segments, but can mean 'frugal engineering' in some instances (Nathan and Sarkar 2013: 4–12). In the Global South, large segments of the commodity markets are characterised by high volume and low value; this is unrelated to the highest global quality standards or innovativeness. Huf Hülsbeck and Fürst underlines this for the automobile component suppliers. Another example is Bosch's automotive division, which had to redesign their electronic control unit for the car model Nano produced by Tata. Usually Bosch develops and adapts products in premium automotive segments, which is a clear high-road strategy. In the case of Nano, top management's goal was to bring down expenses by about two-thirds. In this project, R&D managers had the shared vision that it was necessary to change from a high-road to a low-road strategy and start up 'frugal' R&D (Freiberg, Freiberg and Dunston 2011: 154, Nathan and Sarkar 2013: 13). Multinational companies such as Huf Hülsbeck and Fürst and Bosch therefore have an international division of competencies: one section deals with high-road R&D for the world market and the global product platforms, and the other with adapted 'frugal' R&D products for particular countries of the Global South.

Roland Berger (2013b) defines 'f-r-u-g-a-l' as functional, robust, user-friendly, growing, affordable and local. Apart from the automotive and electronics value chain, the pharmaceutical industry is also engaged in frugal engineering, particularly in India, China and Indonesia. In such countries, not only Western multinationals but also domestic companies are successful in frugal engineering. Sometimes they show significant growth and even succeed in conquering international markets, for example the generic drugs market (Nathan and Sarkar 2013: 11–12).

Particularities of the Product Development Process and Local Problem Solving

Shared visions of the internationalisation of R&D are not only influenced by product markets and the labour force. Another important factor is the particular organisational setup within the multinational company. Organisation of the product development process and local problem solving are important reasons for establishing engineering competencies abroad. Such internationalisation of R&D-related knowledge is widespread and often occurs step by step (Fuchs 2003a, 2003b, 2005). Nevertheless, these changes in the international division of R&D-related knowledge are often overlooked in the media as they are less visible than the 'global race for talent'.

Internationalisation of knowledge can become necessary during the 'front end' of a new product development process; that is the period immediately after a new project idea is suggested (Kijkuit and van den Ende 2010: 451). Depending on the location of the new project, international exchange of knowledge can be necessary, for example resulting from common networking and 'a little help from our colleagues' (Kijkuit and van den Ende 2010: 451). Blomqvist et al. (2004: 592) emphasise intra-organisational and international collaboration as a 'meta-capability' for R&D. The initial innovation period requires a large amount of coordination, team meetings and contacts, building on hierarchical patterns and hierarchy-free spaces (Pearce 2004: 47, Persson and Åhlström 2013: 68–70). In such early stages of innovation the capability to unite the different shared visions of the various team members is particularly important (Song et al. 2007).

During the later stages of innovation, shared visions between the different units remain important (Kiella and Golhar 1997: 197). Shared visions are relevant for communities of practice whose members act in globally distributed projects (Kahn and McDonough 1997: 51). Such communities of engineers, technicians and other professionals need shared cognitive and interpretative contexts for their collaboration. Going beyond the obvious such as speaking the same technical language (Huber 2012), these collaborative contexts can relate to a common educational background (such as specific universities), common academic disciplines, and particularly also shared experiences with regard to the company, familiarity with specific work tasks and specific expertise (Gertler 2008: 203, 209–10).

Bridging different socio-cultural contexts needs time as shared knowledge and shared visions can only be learnt through practice. An investigation by McKinsey (2011b) reveals that it is still difficult for many companies to share knowledge internationally; two-thirds of the respondents stated international knowledge sharing in their companies was no better than 'somewhat effective'. Apart from telephone and video conferences (65 per cent of the respondents), face-to-face meetings are considered highly relevant (62 per cent). Obviously travelling and face-to-face contacts are important channels for developing mutual knowledge and shared visions between different socio-cultural contexts, besides IT and electronic media.

In R&D such requirements for international cooperation and collaboration have increased during the last 50 years. The organisation of R&D has changed profoundly in many companies, developing from linear organisation of innovation (from the invention of a new product to production and diffusion in the market) towards interactive organisation with many feedback loops in the process, including permanently changing market requirements. 'Simultaneous engineering' describes the management of the diverse actors taking part in this process and the coordination of particular product development tasks, often at an international level. Apart from the feedback loops in the company, there is also closer multiscalar cooperation between companies, universities and further research institutes (Westlund and Li 2013: 130). Innovation is part of, and the result of an interactive, often non-linear process involving different actors. It is a principal task for management to organise such knowledge networks (Martin 2012a: 11, 2012b: 1579).

This is particularly true for large, multinational companies. Already Bartlett and Ghoshal (1986) suggested to 'tap your subsidiaries for global reach', proposing to decentralise R&D for the purpose of market expansion. Today, coordination between central and decentralised R&D and decisions regarding the outsourcing and coordination of R&D with external partners is daily business in many multinational companies (Edler, Meyer-Krahmer and Reger 2002: 151, 162). Given the scope and complexity of R&D in multinational companies and global value chains, innovation is strongly influenced by the ability of a company to use its internal knowledge and combine it with external knowledge, developing its knowledge strategically, mobilising it and applying it effectively (Hirsch-Kreinsen 2008: 28–9). 'Open innovation' (Chesbrough 2003) is the word of the moment. Obtaining access to external knowledge means networking, which can take the form of co-development partnerships and outsourcing of innovation capabilities at different spatial levels from the local to the international.

The concept of R&D as an open learning system implies that the company must combine emergent, unplanned knowledge with deliberate knowledge (Belussi and Sedita 2012: 166–9). Scientific-technical knowledge needs to be integrated with other kinds of knowledge (Knoben and Oerlemans 2012). Loops of both scientific-technical knowledge and experiences are considered important for innovation, including a significant delegation of responsibilities (Arundel et al. 2007: 1176). Recently, the uncontrolled and spontaneous elements of the innovation process have come into particular focus (Tödtling, Prud'homme van Reine and Dörhöfer 2011: 1886, Chen, Chang and Hung 2008). Innovative regions are thus considered to be open learning systems, combining 'cross-over' emergent, unplanned knowledge with deliberate knowledge (Belussi and Sedita 2012: 166–9).

In R&D, international communication, coordination and fine-tuning have obviously become more and more important during the last decades. This is mainly caused by changes in the organisation of R&D and the growth and globalisation of R&D departments.

Apart from knowledge sharing between international R&D sites in the product development process, R&D can also grow at peripheral sites *because of problems on the shop floor*. Many engineering activities are decentralised across the world, because they are closely related to the internationally localised manufacturing process. This is because production cannot be planned from A to Z, with top down knowledge transfer from the parent company to the production site abroad. A lot of unexpected and sudden difficulties can occur in foreign plants, which require local problem solving and at least a minimum of R&D.

Local problem solving can imply the development of new production technology, such as large engines and specific machines, and relate to a broad range of equipment and tools. R&D-related activities abroad can therefore be minor or major, ranging from particular machines or parts to larger systemic technologies or processes in the whole plant. Engineering also comprises aspects such as job design and work ergonomics, including environmental requirements and the change towards 'green' innovation (NIST 2010). Hence, there is close interaction between product and process-related research, design, construction and engineering activities and the work organisation on the shop floor, with numerous feedback loops to eliminate flaws. Put another way, manufacturing abroad needs R&D services as a form of support. The support of production by R&D-related activities is an example for the globalising 'servitization' of manufacturing (Sundbo and Toivonen 2011: 2).

Often, modularisation is considered in order to overcome the particular difficulties of the international product development process and reduce problems abroad. In general, modularisation reduces complexity by standardising interfaces, which simplifies and accelerates communication and coordination. Modularisation streamlines the product development process and allows a broad range of product variants to be created, resulting in product families with different models for the different market segments. Modularisation also accelerates the development of new product generations (Persson and Åhlström 2013: 56). During the last decades, such formalisation has become common through the worldwide implementation of IT systems, in particular computer aided design (CAD).

Modularisation supports coordination across distances and as such is highly relevant for the internationalisation of R&D. Initially, this type of internationalisation implied a trend towards globalising 'D without R', in other words the internationalisation of less sophisticated tasks and hence the relocation of mainly applied design and adaptive utilities. The less complex design tasks are those which are easy to standardise, codify and thus transfer in modules across borders; they were the first to go global (D'Agostino, Laursen and Santangelo 2013).

The electronic industry was one of the first to use modularisation as a basis for re-organising the product development process. Strong global suppliers of computer software delivered turnkey solutions to their clients and thus restructured the worldwide value chains and knowledge networks (Sturgeon 2003: 199–203). During the last decades, modularisation has spread to large parts of international product development processes in diverse industries including the automotive industry (Freyssenet and Lung 2004). It has even arrived in the mechanical engineering industries as 'flexible standardisation' (Hirsch-Kreinsen 2009: 5–9).

Despite its advantages, modularisation cannot overcome all the difficulties experienced in international R&D departments. More complex R&D has therefore also begun to internationalise. Yang and Chen (2013: 69–71) illustrate the internationalisation and modularisation of R&D-related knowledge in the notebook PC industry in Taiwan and the People's Republic of China. They recognise a high degree of informal technical collaboration and a considerable exchange of tacit knowledge among brand-name companies, the key component suppliers and original design manufacturers at the international level. This is notable since the electronics industry is generally emblematic for modularisation and thus for few coordination requirements on the international level (Sturgeon 2003).

Unexpected problems will always occur somewhere in the product development process. Since these require individual solutions, McKinsey (2012c) recommends to managers: 'Don't standardize more than is necessary. ... Prefer standard principles to detailed rules for local processes'. Managers should only use a set of global principles and 'not chapter and verse' (McKinsey 2012c).

Overall, international patterns of R&D are becoming increasingly complex: apart from standardised knowledge, informal knowledge has also begun to migrate worldwide since additional exchange of knowledge is necessary beyond the well-defined interfaces in the global R&D networks. Knowledge and interpretation are shared internationally up and beyond modularisation. The traditional view is that modularisation has caused a division between 'R' in the core economies and applied 'D' in the subsidiaries abroad, but it is clear that diverse kinds of knowledge have now become merged. Chapter 6 discussed the multiple and segmented multinational organisation and the resulting need for fine-tuning

(Segrestin, Lefebvre and Weil 2002: 68). Given such diversity, modularisation cannot solve all the problems in the various international sites.

Localised Interaction with Actor Networks and Institutions to Promote R&D

Up and above the factors outlined above, localised interaction with actor networks and institutions to promote R&D in overseas regions also plays a role in the globalisation of knowledge. Regional settings are important in managements' shared visions on the globalisation of knowledge-intensive tasks. Therefore, multinationals' subsidiaries and R&D sites are increasingly integrated in global–local relationships. They are part of multiscalar networks which extend from the local (the location of the subsidiary) to the national and international level. The subsidiary can thus be said to bundle multiscalar relations. The effects of the global–local interplay on localised knowledge creation are well-documented in the academic literature (Asheim and Isaksen 2002, Bathelt, Malmberg and Maskell 2004, Gertler 2008, Lagendijk 2001, Oinas 2000). This section focuses on the location of the subsidiary and the localised interaction of the subsidiaries with multiscalar actor networks and institutions promoting R&D.

The academic literature describes the destinations of knowledge-intensive investments as 'centres of excellence', 'pockets of excellence', 'hotspots' of R&D-related knowledge (Manning, Sydow and Windeler 2012: 1203), 'platforms of innovation' (Cooke et al. 2010) and 'knowledge and innovative hubs' (de Propris and Hamdouch 2013: 997). Integrated in global networks of knowledge and production, these are places of internationally integrated R&D networks and global innovation. They also form part of international scientific networks (Wyckoff 2013: 314). In the core economies, certain areas of New York, London, Paris, Montreal and Los Angeles represent such hotspots; in the emerging countries it is Shanghai, Beijing, Bangalore, Manila and others (see Chapter 7).

However, metaphors such as 'hotspots' should not overvalue the well-known metropolitan areas in the mature core economies and emerging markets. As McKinsey (2011b) noted in a survey with executives, managers regarded Shanghai and Beijing as the most important regional choices for R&D destinations in the emerging economies. Notably, they also considered other places in the peripheries of the world economy to be attractive. Many of these places are less popular than the well-known R&D locations in North America, Europe and Japan, or even in China and India, indicating that a considerable part of R&D in the advanced manufacturing industries is internationalised in 'hidden' places or 'hidden' hubs. Some of these hubs are located near large metropolitan areas, others have developed over time at particular places through the growth of their industries (see McKinsey 2013: 16, 42). Thus, many places are attractive on account of their proximity to new markets *and* highly qualified workers *and* relatively low labour costs; they benefit from agglomeration economies. Additionally, such locations often offer initiatives for further education and training and provide organisational and institutional support. Such locations are neither centres of excellence nor simply extended workbenches. They offer attractive conditions to R&D and production and integrate various stages of the product development process and levels of competencies.

There are several cases which illustrate the localised global-local interplay. Lema, Quadros and Schmitz (2012: 11–17) highlight the case of Brazil where there are joint activities of the various subsidiaries of multinational companies, their direct first-tier suppliers, universities and research centres. Today the Brazilian automotive industry is deeply involved in the innovation process by co-design. The relations between the Brazilian firms and their foreign partners are becoming denser and more bi-directional. In the 1990s for example, the German automobile component supplier Mahle acquired the Brazilian firms Metal Leve and Cofab, both of which had sizeable R&D departments. On this basis, Mahle established its Brazilian Technological Centre, which is one of five global R&D centres of the company. Presently, Mahle leads a research consortium which seeks to produce new knowledge on flexfuel engines. The consortium includes international automotive firms as well as Brazilian energy companies and research groups at Brazilian universities.

Fernandez-Stark, Bamber and Gereffi (2013: 534–48) discuss examples of local integration with regard to service-offshoring in Latin America, extending the focus beyond just R&D. Multinational companies with headquarters in the mature core economies have begun to offshore their services, particularly in Argentina, Brazil, Chile, Colombia, Costa Rica and Mexico. Global players from India and the Philippines have also invested in Latin America, turning Brazil and Mexico into strong players in the knowledge-intensive service industries. Apart from market size, such developments often hinge on education and further training offered by regional and national actors and pre-existing industrialised pathways in the regions. In Chile, for example, there is particular support for recruitment and vocational training for a highly skilled workforce and thus for the 'high-road' labour market segment (Fernandez-Stark, Bamber and Gereffi 2013: 547).

Localised interaction with actor networks and institutions includes activities as diverse as tax subsidies, public-private partnerships and an advanced research infrastructure. The regulation of intellectual property rights can also play a role. Some governments of the Global South provide R&D-specific pre-investment support, implementation services and after-care services, together with a broad range of activities by ministries, promotional agencies, inward investment agencies and development organisations (Fernandez-Stark, Bamber and Gereffi 2013: 547, Guimón 2008: 4). Such initiatives also attempt to improve the image of the country as an R&D-intensive location, as exemplified by the slogan 'Brazil in the world – Innovative, sustainable, competitive' (APEX 2013, Guimón 2008: 7). Thus, local, regional and national governments work together with regional development agencies and other initiatives to promote knowledge-intensive regional development and play an active part in the localised interaction with multinationals (Figure 11.1).

---

p e r i p h e r i e s   o f   t h e   w o r l d   e c o n o m y

e. g. in regions of Latin American countries

subsidiaries ⟸⟹ ***public and further localised actors*** *at the destination of the subsidiary, framed by the local, national and international institutional setting*

active local management

- *education and further training*
- *support for recruitment of skilled personnel*
- *public-private partnerships*
- *research infrastructure*
- *enhancements in regulation of property rights*
- *tax subsidiaries*
- *particular support for 'high-value' segments of R&D*
- *R&D-specific support in pre-investment, implementation, after-care services*
- *promotional initiatives for R&D and innovation*

(based on Fernandez-Stark, Bamber and Gereffi 2013: 547, Guimón 2008: 4)

Figure 11.1   Examples for the promotion of R&D-intensive regional development in Latin America

In the subsidiary, active local management plays an important part in facilitating integration into local actor networks and institutional arrangements overseas (Fuchs 2003a, 2003b, 2005). Local management has the difficult task of coordinating the diverse and partially contradictory demands of headquarters, the subsidiary and other local actors, often in a 'sandwich position' between foreign headquarters and the local workers (Girndt 2012: 23, Molitor 2012: 33). Sometimes, local managers are domestic, lending them a stronger 'local perspective', but more often they are expatriates sent by the parent company. McKinsey (2011a) reveals that only one third of R&D projects is managed by locals, implying that R&D managers are expatriates in two thirds of the cases. Some multinationals train domestic managers through a 'reverse expat' strategy, where managers who have grown up in one country gather experience in locations abroad before returning to their domestic plant, thus developing a broader 'global–local' experience (McKinsey 2011a).

Destined to bridge boundaries (Glückler 2011: 222), such expatriates perform in a heterogeneous field of different roles and self-perceptions, acting as 'inspectors', 'missionaries', 'global managers', 'citizens of the world' – or as 'foreigners in the local subsidiary' (Wagner and Vormbusch 2010: 200–229). Local managers also need to account for the politics and power relations in the global corporation which is a geographically, socially, culturally and institutionally differentiated organisation (Geppert and Dörrenbächer 2011: 6).

Hence, headquarters are not the only ones steering the gradual process of regional subsidiary integration, as local management plays a prominent role in subsidiaries becoming locals. Academic research should include such bottom-up perspectives (Williams and Geppert 2011: 73) and embrace the particular settings of headquarters–subsidiary relationships (Blazejewski and Becker-Ritterspach 2011: 177). Local management does not simply receive orders from the top. There are many ways in which local executives can respond to headquarters' requests (Schotter and Beamish 2011).

Apart from the self-definition and engagement of active players in the region and for the benefit of the region, money is the other more tangible aspect. Innovative places are 'made' or at least supported by funding, including public funding from national or local government as well as the funding provided by

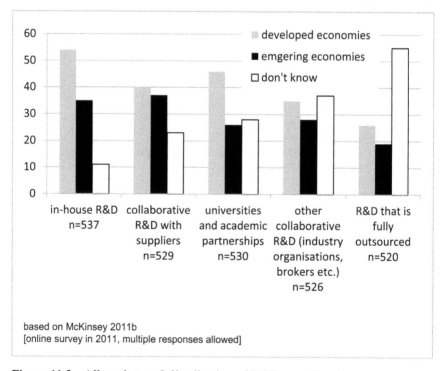

based on McKinsey 2011b
[online survey in 2011, multiple responses allowed]

Figure 11.2   Allocation and distribution of R&D spending in company budgets in 2011, in per cent of respondents

the multinationals (Autio, Kanninen and Gustafsson 2008, Arundel, Bordoy and Kanerva 2008: 32). McKinsey (2011b) shows that multinationals invest in in-house R&D as well as partnerships at the destinations of their foreign direct investments. These investments include collaborative R&D with their supply firms, universities and academic partnerships, industry organisations, or brokers. Figure 11.2 suggests that these external relations are particularly relevant for R&D in the emerging economies.

To conclude, as a result of localised interaction of multinationals with actor networks and institutions to promote R&D, at least some peripheries of the world economy are advancing with respect to knowledge-intensive tasks and R&D. The following section will look at such dynamics in more detail. At the micro-level, the most striking consequence is a considerable internationalisation of complex R&D functions, with co-design of 'R *and* D'.

The perspective now shifts away from aspects *influencing* shared visions of management towards the *effects* this is having, in other words how the internationalisation of R&D takes place.

From 'D *without* R' to Co-Design of 'R *and* D' in the Peripheries of the World Economy

R&D internationalisation takes place according to various and changing patterns in multinational companies. Commonly held insights about the localisation of R&D now need to be revised. Up until the early 2000s, multinationals were assumed to concentrate their R&D at headquarters, especially the highly advanced and sophisticated research ('R') for new products or new product components. Only selective functions of 'D' and product adaptations of less complex design were decentralised. In short, the view was that peripheries only have 'D *without* R'. This pattern was thought to result from economic factors, mostly related to employment. Because of the relative cost advantage of engineering in the peripheries, multinational companies tended to relocate simple design and engineering tasks to the peripheries while retaining complex tasks at home (Gomory and Baumol 2013: 24–5). Until the early 2000s there was strong empirical evidence for this, for example from the brand-name producers of the automotive industry and their first-tier suppliers. The shared vision of management was generally that basic product research and development should take place at the parent company, while standardised development and design should be carried out in other countries with lower labour costs.

Currently management tends to think about R&D organisation as a matrix in which different cells or tasks can be internationalised (Deschryvere and Ali-Yrkkö 2013: 180). There is increasing 'organisational decomposition of the innovation process' (Lema, Quadros and Schmitz 2012: 11) which not only applies to simple innovation, but also to complex and sophisticated engineering. This used to be carried out in the parent company but has now been internationalised into worldwide subsidiaries or offshored in the peripheries of the world economy (Lema, Quadros and Schmitz 2012: 11). Thus, since the late 1990s companies have gradually learned how to successfully internationalise new and increasingly complex R&D tasks. Today we find cases of internationalisation of refined R&D and co-design in the cores *and* peripheries of the world economy. Some multinational companies have developed global product platforms which go far beyond minor innovation for local adaptation in the emerging economies (McKinsey 2011b).

Obstacles remain in the internationalisation of complex design tasks. The lack of skilled employees for R&D overseas is a bottleneck, which is confirmed by McKinsey (2011b) who interviewed 1,100 executives of companies. Respondents stated that the performance of R&D managers in emerging economies tends to lag behind that of their colleagues in the core economies; in consequence many R&D decisions are taken centrally. Three quarters of the interviewees commented that criteria for evaluating project portfolios are selected at central offices rather than local sites. Skilled domestic managers are hard to find and harder still to keep in the subsidiary. Particularly in the emerging economies of China, India and Brazil, locals have high expectations with respect to salaries and careers, putting the subsidiaries of multinationals into direct competition with domestic employers (McKinsey 2012a).

This sceptical view of the present situation is in contrast to the respondents' future expectations. Those managers expecting more globalised R&D in their companies assume that talent and organisation overseas will improve (McKinsey 2011b). Beyond R&D, some companies have already relinquished the single-headquarters model and established secondary headquarters, or split head office functions with sites overseas (McKinsey 2013: 16). Cities in emerging countries are expected to play an increasing role as centres of decision-making, including R&D (McKinsey 2013). In 2025 there could be three times as many large company headquarters in emerging countries as in 2010, and the emerging countries could comprise nearly half of the companies in the Fortune Global 500, particularly in regions of China, Latin America, Eastern Europe and large parts of Asia (McKinsey 2013: 13).

Hence, 'D without R' is no longer an appropriate description for the general location principles of R&D-related knowledge in subsidiaries abroad. The core-periphery model is an oversimplified theoretical notion. Currently, the internationalisation of R&D generates mixed patterns of 'D without R' and of 'R and D'. The following case study describes an advanced R&D centre in Mexico which conducts complex 'R and D'.

Internationalisation of R *and* D: An Example of an R&D Centre in a Periphery

Delphi automotive is a prime example for illustrating the internationalisation of R&D into a range of in-house technology centres around the globe. Going far beyond the gradual decentralisation of knowledge into the R&D departments of existing subsidiaries, *Delphi Technical Centers* clearly include 'R and D'. The following describes the case of the R&D centre in Northern Mexico, including a closer look at Delphi's role in the North Mexican *maquiladora* region. The case study was first researched in 2001 (Fuchs 2001, 2003a) and updated in 2013.

Delphi is a leader in automotive technology with nearly 120,000 staff in production and R&D in 32 countries. By 2013, Delphi had established 15 technical centres in 11 countries, employing 18,000 scientists and engineers (Delphi 2013). The most important technical centres have been dispersed across the large market areas of the world in order to serve these markets and be close to the customers (see Map 11.2). A *Delphi Mexico Technical Center* was established in Ciudad Juárez in 1995. Mexico also has two smaller centres in Saltillo and Querétaro (Delphi 2013).

The *Delphi Mexico Technical Center Juárez* is the largest Delphi R&D unit globally, with clients distributed all over the world, especially in North and South America, Europe and East and South-East Asia (Carrillo 2013). In 2013, the site had 2,093 employees, of which 1,403 worked in engineering and 690 in non-engineering. For the period of 1998 to March 2013, the *Delphi Mexico Technical Center Juárez* held the company record of 1,069 inventions. In the same period, it filed 411 patents and had 308 patents granted (Delphi 2013). A wide variety of engineering tasks are carried out at the site, with core competencies in product design, process development, process validation, testing, prototyping and engineering services. The site has testing capabilities and laboratories for material analysis, calibration, vibration and instrumentation, as well as labs to test durability, corrosion, substances of concern, analytical engineering, rapid prototypes and packaging (Delphi 2013; Figure 11.3). The *Delphi Mexico Technical Center in Juárez* thus also serves as a good example for the importance of globalised testing activities.

Since the early 2000s other foreign companies have also established or expanded technology centres in Northern Mexico, which are dedicated to product research and the production process (Carrillo 2013). Siemens Automotive built up engineering centres in the Mexican cities of Guadalajara and Cuautla; a unit which is now owned by Continental Powertrain (Continental Powertrain 2012: 19, 27). Another example is Honeywell's *Mexicali Research & Technology Center* in Baja California which employs about 350 people in the design, engineering and testing of components for aircraft systems (Honeywell 2013).

Generally, the expansion of sophisticated R&D into Northern Mexico is motivated by the proximity to the USA, the relatively cheap labour costs, skilled labour and the long experience of multinationals with efficient production and engineering in Mexico (Carrillo 2013). Managements' shared visions of R&D in Mexico thus result from successful early beginnings in assembly in Mexico, then moving to complex manufacturing and finally engineering with increasingly complex functions. This gradual internationalisation of knowledge is frequently found in the peripheries of large existing market areas (Fuchs 2003a, 2003b, Fuchs and Winter 2008, Winter 2008, 2009). Shared visions of the relevant actors

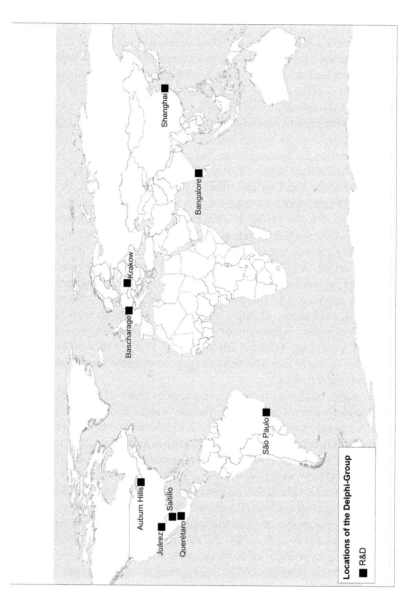

Map 11.2   *Delphi Technical Centers* worldwide 2013
Source: Based on Delphi 2013; cartography: Jonathan Mauersberger.

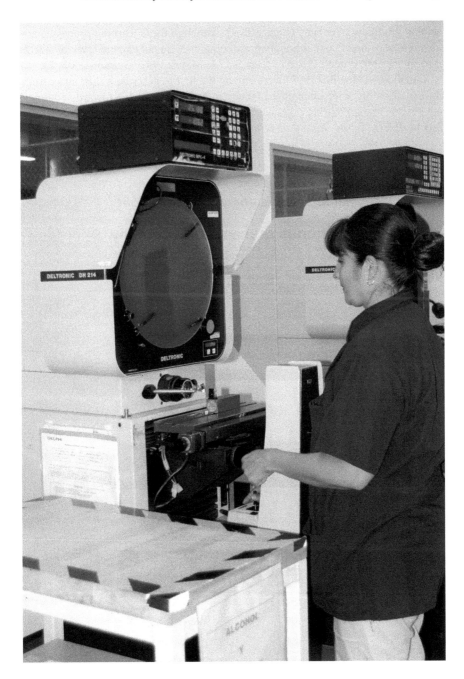

**Figure 11.3   *Delphi Mexico Technical Center* in Ciudad Juárez: Optical
　　　　　　Comparator**
Source: Photo: Delphi, with the kind permission of Delphi Mexico Corporate Affairs.

develop gradually over time (Figure 11.4). Although these learning processes are not a one-way street, and although many companies do not expand towards R&D, gradual learning processes are widespread (de Meyer 1992, Guimón 2008: 2).

Carrillo and Hualde (1999) were among the first to describe a shift away from Fordist assembly with unqualified labour in Northern Mexico towards complex and flexible manufacturing organisation shaped by high quality standards and later also by R&D. While *maquiladora* plants usually merely import parts from the

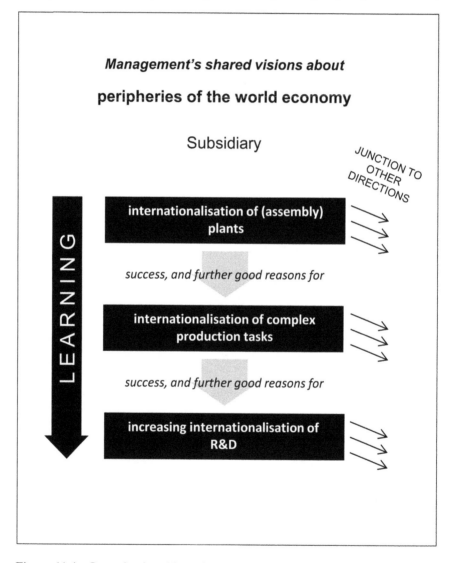

**Figure 11.4   Organisational learning: gradual internationalisation
of knowledge**

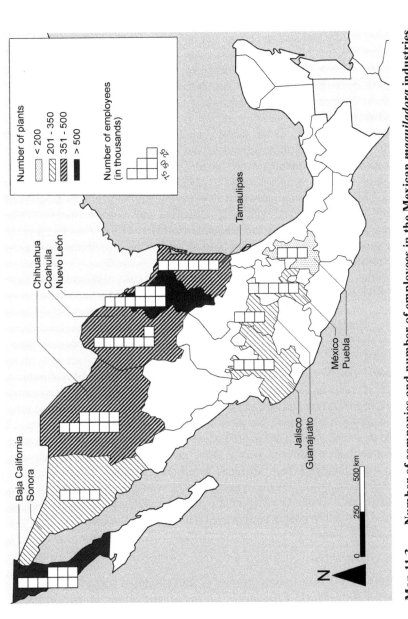

Map 11.3   Number of companies and number of employees in the Mexican *maquiladora* industries
Source: In 2012, based on INEGI 2013, cartography: Jonathan Mauersberger.

USA, assemble them and export the products back to the USA, the authors noted a widespread increase in R&D functions in the *maquiladoras* in the late 1990s, incorporating the development and testing of products as well as production process engineering.

It is noteworthy that such processes not only occur in the large R&D centres, but that many plants have small departments for adaptive product and process development (Fuchs 2003a). The regional importance of such small and 'hidden' engineering departments should not be underestimated in terms of generating employment for engineers and technicians.

Consequently, the *maquiladora* region is multi-faceted. Presently, it still represents an archipelago of the world economy which is integrated in global economic networks and thus dependent on external decisions (Veltz 1996). In this respect, the *maquiladora* can be considered a periphery, reliant on the industries of the core economies and a place of 'desert capitalism' (Kopinak 1996), combining economic growth with the exploitation of human labour and the natural environment, and with drug trafficking and violence (Berndt 2013). Given the rapid expansion of the *maquiladora* (Map 11.3), it is questionable whether such economic development is sustainable. Since 2000, the number of employees has doubled and now amounts to about 2.5 million people (INEGI 2013). Workers receive an average monthly wage of about 8,500 Pesos (500 Euro or 650 US-Dollar, INEGI 2013). As the OECD states, in the Mexican *maquiladora* educational attainment levels remain considerably below those of the US side of the border. However, the region is improving rapidly with respect to vocational education and further training, considerably faster than many other Mexican states (OECD 2010).

To conclude, Northern Mexico can be considered a multi-layered region. Although it remains a periphery of global manufacturing, it does show considerable trends towards upgrading and R&D. The *maquiladoras* benefit from internationalised R&D which has positive impacts on the region; Chapter 12 returns to this topic with regard to production-related knowledge. However, huge social and ecological problems persist. Also, many companies remain assembly plants at heart that have been upgraded to different degrees. The dynamics of Ciudad Juárez cannot change the dynamics of the whole *maquiladora* region, let alone Mexico as a whole. Once again, the container metaphor of places needs to be replaced by a differentiated perspective on knowledge, the economy and social development in space.

Indications for the Internationalisation of R&D-Related Knowledge into Peripheries of the World Economy

Similar processes of internationalisation of R&D-related knowledge are occurring in other regions of the world. The following highlights some of the current trends as evident in selected sectors of the manufacturing industry and at specific locations.

Automotive Industry

The automotive industry is showing trends of internationally extending R&D. Although the leading brand name car producers still do most of their basic research at home, internationalisation of engineering activities is increasingly common. In line with the strategy of building on common platforms of design and development, the Volkswagen Group keeps relevant R&D capacities centralised, but has dedicated R&D centres for the different brands of the Volkswagen Group (Hauser-Ditz et al. 2010: 129–30). Furthermore, there are international R&D cooperation projects, such as the New Beetle which was designed collaboratively by Puebla in Mexico and the parent company in Wolfsburg in Germany (Fuchs and Endres 2007). In 2012, the Volkswagen Group employed more than 5,000 staff in engineering in North and South America and Asia compared to nearly 10,000 in engineering at headquarters in Germany (Volkswagen 2013a). Similar patterns can be found at Daimler. The core of R&D is in Untertürkheim (Germany); yet each division of the corporation is responsible for the respective product development process (Hauser-Ditz et al. 2010: 92). Another example is Renault's investment in Dacia (Romania): Renault not only invested in production, but also built a large R&D centre in Romania, building up further engineering capacities in Bucharest, Pitesti and Titu (Pavlinek 2012: 294–6, 302). This is in line with Renault's strategy to locate R&D in different parts of the world. Apart from Romania, there are locations in Spain, Brazil, India and South Korea, with a strong strategic R&D centre in France (Hauser-Ditz et al. 2010: 213). Many first-tier suppliers follow suit and also internationalise their R&D.

This can be described as a change of the 'productive models' in the automotive industry. Formerly, such productive models were largely based on the European, US and Japanese automotive industry, with particular company product policies, company organisation, labour relations and links to the public sector in the advanced core economies (Freyssenet and Lung 2004). Today, productive models have become amplified and internationalised. They include a broad range of worldwide locations. Furthermore, there are new competitors as well as cooperation partners to account for, in particular the new domestic automotive industries in the emerging economies such as India and China (Jullien and Pardi 2013: 97–8).

Consequently, some emerging economies show considerable new R&D capacities in automotive engineering. Lema, Quadros and Schmitz (2012) describe the case of Brazil, which initially only had process engineering as a way of meeting the quality standards of world market production and also local demand. Innovation at this stage was limited to adapting foreign platforms to the Brazilian requirements ('tropicalisation'). Later the range of tasks expanded and included the design of complete derivate vehicles with new options and various new items. Today, the four leading companies of General Motors, Volkswagen, Fiat and Ford have built up strong product innovation capacities and allocated particular product development tasks to Brazil. The number of car models developed in Brazil is increasing. Recently, the Brazilian companies gained innovative competencies not only for the home market, but began to develop design and engineering in the lead firm's global innovation network. This awarded them competencies as platforms of innovation (Lema, Quadros and Schmitz 2012: 19–20, 33–42).

These changes in the Brazilian companies are not accidental. Brazil is a 'big emerging market' and thus highly attractive to investors. Also, there is considerable public support for ventures. Last not least, domestic companies have also invested in skilled employees, modern equipment, efficient organisation and external relations, enabling them to link up with the global value chains (Lema, Quadros and Schmitz 2012: 17).

Electronic Industries, Software and Equipment

Electronics, including information and communication technology, partially supply the automotive industries and thus follow the trends described above. Other parts of the electronics industry supply other segments of producer and consumer markets. With respect to assembly, the electronic industries went global early and comprehensively, a trend later duplicated in R&D. The Mexican *maquiladora* industry is an important and dynamic location for such activities in America; the Chinese Guangdong province is the Eastern Asian equivalent. Like the Mexican *maquiladora*, the Guangdong province has recently developed into a flourishing region, although it continues to be shaped by high external dependency and thus by international up and downturns (Schiller 2013). At the same time, R&D is expanding. Private R&D investment is particularly important for the Guangdong province as foreign direct investment plays a crucial role in the increase of R&D-related activities. Again, like other increasingly knowledge-oriented regions, the Guangdong province has an active public sector, with local government actively improving the links between science, technology, research, applied development and the public sector and making available pilot locations for science and technology (Kroll and Tagscherer 2009: 3–8, 13–21).

In India, another big emerging market, R&D is expanding particularly in the software industry. Lema, Quadros and Schmitz (2012: 20–21, 36–40) show that today, Indian software firms are increasingly competent in process and organisational innovation, including problem identification and problem solving. Thus, there is a tendency for upgrading from low-level design via architectural design to global product responsibility. Similarly, Parthasarathy (2013: 382) argues that the Indian software industries show a broad range of skill intensities, ranging from rather simple business process outsourcing via engineering services and knowledge-process outsourcing to complex R&D.

The PC industry is closely related to electronics and internationalised early. Taiwan in particular had many global lead firm branch offices which cooperated with local suppliers some decades ago. Engineering activities gradually also expanded, and R&D centres have been built up there since the early 2000s for the notebook industry, for example by Dell and HP. Cooperation in R&D has been extended, giving rise to co-design (Yang and Chen 2013: 73–4). The Taiwanese

IT industry is expanding to the People's Republic of China, with collaborative adaptation involving numerous interconnected firms and orchestrated by a variety of supporting initiatives and institutions (Lee and Saxenian 2008: 173).

In the chip industry, a shift from the core economies to Asian countries has been occurring for some decades, with R&D capabilities established in Korea, Taiwan, Malaysia and Singapore (Nathan and Sarkar 2013: 5–7).

Mechanical Engineering Industry

An increasing amount of mechanical engineering R&D is conducted overseas. Proximity to clients abroad is the key driver here, with low-cost production generally less important than in the electronic industries which are quite sensitive to labour costs in engineering. In mechanical engineering, it is single orders and the specific demands of small series production which require a particular amount of R&D abroad at the client's location. Knowledge transfer in these cases is not limited to production as mechanical engineering companies not only sell finished products such as automata, robots and large plants, but also the know-how required for running the machine correctly, maintaining and repairing it. Rather than a mere by-product, these knowledge-intensive services form an essential part of the core business of these industries. Often, such knowledge is further developed in close interaction with the client, offering local engineers and technicians the opportunity to gain insights via joint engineering activities (Fuchs and Kempermann 2012, Scholz 2013).

Pharmaceutical Industry

The pharmaceutical industries can serve as another instructive example for R&D internationalisation. In the pharmaceutical industries, R&D is highly internationalised in worldwide R&D networks. Global R&D sites were given equal rights and responsibilities early on, making them partners in the R&D network. Many research projects are integrated in multi-site projects, and often, each R&D site has a special leading role as a competence centre for a particular product development process (Gassmann and von Zedtwitz 1999: 244).

Recent pharmaceutical R&D internationalisation has extended to companies with headquarters in the mature core economies as well as the emerging countries such as India. Haakonsson, Ørberg Jensen and Mudambi (2013: 681–95) describe the growth of highly advanced R&D in India since the mid-2000s, which was enabled by the improvements of intellectual property rights in India. The new regulatory framework attracted much investment by pharmaceutical companies of the core economies. At the same time, the domestic Indian pharmaceutical industries expanded their own R&D, particularly in line with 'frugal engineering', to become global players.

To conclude, internationalisation of R&D is progressing apace in different industrial sectors, supported by public and private engagement and the development of suitable institutional frameworks.

The perspective now moves away from the micro-level – the perspective on companies and the locations of their subsidiaries – to the macro-level. Are the processes on the micro-level replicated on the macro-level?

Indications for Change on the Macro-Level

The above section concluded that R&D locations are no longer confined to the traditional core economies. Notable R&D locations have now sprung up in some big emerging markets and also in the peripheries of large existing market areas. But can more general trends be recognised with respect to R&D internationalisation at a country level? The following sets out four major trends: first, the classic global division between the core economies and the peripheries of the world continues to persist. The USA, Japan and some European countries still play a major role at the macro level, with Korea acting as an 'old new' performer. Second, a closer look at the core economies reveals that some countries, in particular Nordic countries, have reached strong positions. Third, some big emerging markets such as China represent new players at a macro level. Fourth, some peripheries of large existing market areas have caught up, especially those which have recently joined the European Union.

Despite the plethora of case studies, statistical information on the internationalisation of R&D is severely limited (Tomiura, Ito and Wakasugi 2013: 246). The following is therefore based on OECD data (2013a) which classes R&D as basic research, applied research and experimental development. The OECD uses expenditure on R&D, patents and researchers as key indicators for the knowledge intensity and innovativeness of countries. These indicators clearly have limits. For instance, they cannot do justice to more comprehensive perspectives on research (see Chapters 3 and 4); also, the range of available country data is selective. Still, the indicators do highlight patterns and trends with respect to different countries' R&D capacities. The statistical information leaves the question open whether R&D supports the performance of a company or whether a high company performance induces more R&D expenditures.

Expenditure on R&D

The main indicator used for international comparison is the gross domestic expenditure on R&D. This is the total expenditure on R&D by domestic companies, public and private research institutes and universities. The indicator reflects the strong position of the mature core economies, but also shows the strength of China (Figure 11.5). In terms of the share each country contributed to the total R&D expenditure of OECD countries in 2009, the United States are ranked top

(42 per cent of the total OECD gross expenditure on R&D), followed by – but trailing considerably – Japan (15 per cent) and Germany (9 per cent). These figures obviously also reflect the size and the economic strength of the respective countries (OECD 2013a, 2013e).

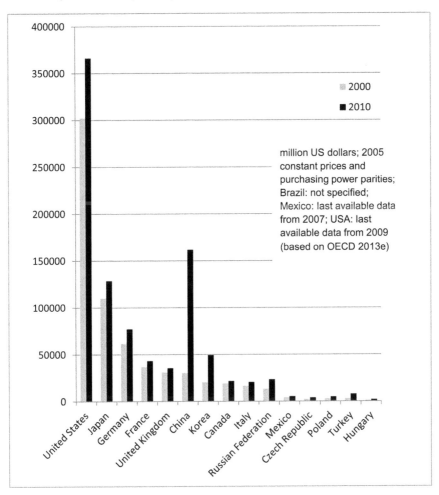

Figure 11.5   Gross domestic expenditure on R&D in selected countries 2000 and 2010

The share of R&D in the gross domestic product is a more detailed indicator. On average, R&D made up 2.4 per cent of the gross domestic product of the OECD countries in 2009. Denmark, Finland, Israel, Japan, Korea and Sweden were above average. In the first decade of the 2000s, Estonia, Korea, Portugal and Turkey were the fastest growing countries with respect to R&D intensity. However, Estonia and

Portugal are small countries, and thus the absolute expenditure is too small for these countries to appear in Figure 11.5.

In China the average annual real growth of R&D spending has been almost 20 per cent. Thus, China has recently become the largest contributor to R&D investments worldwide besides the USA. From 2000 to 2010, the share of R&D in the gross domestic product almost doubled in China, increasing from 0.90 per cent to 1.77 per cent (OECD 2013a).

Thus, some similar dynamics can be noted for selected big emerging markets and peripheries of large existing market areas. These, however, should not gloss over the huge worldwide disparities. As the OECD (2012a: 37) states, the share of R&D expenditure in the gross domestic product was 2.32 per cent in the developed countries, 1.11 per cent in the developing countries and 0.20 per cent in the least developed countries (in 2009).

Patents

Patents are an indicator of the inventive output of a country (Figure 11.6). As simple counts of patents filed at national patent offices only have limited international comparability, the OECD uses the indicator of 'triadic patent families', which are defined as sets of patents registered in the different countries to protect the same invention. The USA holds 28.1 per cent of all patent families (2010). Relating patent families to population size shows high values for Japan, Switzerland, Sweden and Germany. In the last decade, patent families have begun to more readily originate in Asian countries (OECD 2013c, 2013g). Once again, the well-known pattern of economically strong and powerful countries is confirmed by patent-related indicators, including the dynamics of the booming Asian countries.

Researchers

Another indicator is the share of researchers out of the total number of employees in a state. The OECD defines researchers as professionals engaged in the conception and creation of new knowledge, products, processes, methods and systems, as well as those who are directly involved in the management of research projects. In the OECD area, 0.76 per cent of the employed are researchers (in 2007).

Generally, two-thirds of all researchers are engaged in the private sector in the OECD countries (in 2007). In the USA four out of five researchers are employed in the private sector and about three out of four in Japan and Korea. In contrast, less than half of the EU researchers work in the private sector and more than the half in the public sector (OECD 2013b).

Remarkably, the Nordic countries of Finland and Iceland have the highest share of researchers. Finland is top of the list with 1.7 per cent of all employees (2010). Among the remaining countries, the highest shares are in Korea (1.11 per cent), which is ahead of China by a considerable margin (Figure 11.7). At the same time, the figure also reveals that many smaller countries in the European periphery of

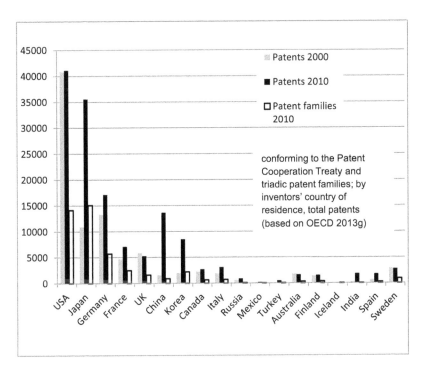

Figure 11.6   Patents in selected countries 2000 and 2010

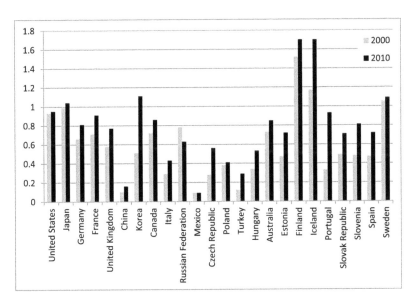

Figure 11.7   Researchers per hundred employed people in selected countries 2000 and 2010

large existing market areas, such as the Czech Republic, Slovak Republic, Poland, Hungary, Slovenia and Estonia, have relatively high dynamics with regard to the share of researchers. Spain and Portugal, which were strongly hit by the economic crisis, also remained in a good position in 2010 (OECD 2013d, 2013f).

The available macro level data reflect dynamic developments as well as persistent patterns. The indicators show that some countries have successfully expanded R&D and that the core economies have remained in a strong position. Rather than a 'global knowledge society', however, R&D is increasing in *selected* regions worldwide.

Interim Summary: Globalisation of Scientific-Technical, R&D-Related Knowledge

The above indicates a general trend towards R&D internationalisation into foreign plants in some peripheries of the world economy. As exemplified by the case of Huf Hülsbeck and Fürst, this is driven by global markets, specific clients' requirements as well as labour costs for engineers. The particularities of the production process and local problem solving were also described as important driving forces. Last not least, the various localised relations of international subsidiaries with actors and institutions also play a considerable role in accelerating the globalisation of R&D. As a consequence, we find numerous examples of subsidiaries in peripheries of the world economy which show a change in the international division of knowledge, moving away from 'D without R' to co-design and even 'R *and* D'. One of the most exceptional examples is the case of Delphi in Ciudad Juárez in Mexico. However, such processes do not simply translate into general macro-trends of R&D internationalisation or substantial effects for the Global South.

To sum up, there are no inevitable effects of R&D globalisation which lead to regional growth, nor is there a story of permanent dependency and inevitable exploitation of the Global South. Rather, the academic literature and the author's own case studies illustrate contradictory trends, with dynamic, complex processes and windows of opportunity at least for some of the locations involved. A lesson is that multinationals and regional actors can act *jointly* to increase R&D in selected locations in peripheries of the world economy as long as certain framework conditions are met and windows of opportunity arise. The following chapter is concerned with the globalisation of production-related knowledge; it underlines such insights from the perspective of blue-collar work on the shop floor.

Chapter 12
Globalisation of Production-Related Knowledge

Moving away from scientific-technical knowledge in R&D, this chapter is concerned with the production process. The last decades have taught multinational companies not only to globalise knowledge in R&D, but also in the manufacturing process. In the large metalworking sector, such trends of gradual learning and upgrading are particularly evident. There, the subsidiaries of multinational companies can become 'integrated production sites'.

Similar to the previous chapter, the following starts with managements' perception of the product and labour markets and the resulting globalisation of knowledge. The relevance of technical and organisational particularities and local problem solving in foreign production sites is illustrated next. The relevance of education, in particular vocational and further training are also emphasised. The chapter then presents the 'integrated production site' as a form of technological–organisational and regional integration. A key quality of the 'integrated production site' is 'upgrading', which is well-known in the academic debate on global value chains (Gereffi, Humphrey and Sturgeon 2005). An 'integrated production site' can encompass product innovation, process innovation, extended and deeper tasks in production; it can also include localised interaction with regional and multiscalar actor networks and institutions. The following underlines and explains the key argument that multinational companies have learnt to globalise knowledge. Selected production plants have the opportunity to upgrade towards 'integrated production sites'.

The chapter then asks how labour can regulate and influence such upgrading processes. International labour regulation and labour relations are often assumed to influence the internationalisation of production-related knowledge as they are taken as stimuli for upgrading in foreign subsidiaries. Constellations are identified in which worker representatives influence upgrading processes at the international level. Only few cases exist where this has taken place, but they are interesting and therefore worth reporting. In this respect, Chapters 11 and 12 are not on a par as issues of regulation and trade union activities are even rarer in international R&D than on the shop floor.

The chapter then looks at the 'objectified' effects of the internationalisation of production-related knowledge. On the micro-level of the firm, the chapter discusses upgrading to integrated production plants as one trajectory and downgrading as another. Macro-level developments are opaque since there is little available information on the internationalisation of production-related knowledge.

Global Clients and Global Labour

The role of global clients and labour in the internationalisation of knowledge has been known for some time (Johanson and Vahlne 1977: 31). Multinational companies began to produce globally in the 1960s and 1970s, generating a division of production-related knowledge which Lipietz (1986, 1987) characterised as 'peripheral Fordism': while highly skilled work was centralised in the parent companies of the advanced core economies, repetitive and monotonous tasks were assigned to the export-oriented production plants of the peripheries. This traditional division of competencies has been discussed from a conceptual point of view in Chapter 7 in the context of spatial dependencies and interdependencies.

Nowadays, there is recognition of changing patterns in the global division of knowledge. *Global clients* are seen as substantial driving forces for subsidiary upgrading. The potential inherent in upgrading is being highlighted both in the academic literature and in the media, reversing Lipietz (1986, 1987) who argued that human labour is particularly exploited in those subsidiaries producing for export. The new view on the beneficial effects of export production has been pushed by the fact there is elbow room for different modes of global production, as well as the debate on the governance of global value chains and the influence of different governance models on upgrading.

Global value chains are ruled by different patterns of governance. As was the case for R&D (see Chapter 11), they all may encompass upgrading. In 'captive value chains', which have a strong lead firm, upgrading can result from clients' requirements. If the lead firm requires knowledge abroad, the supplier has to follow suit. 'Relational value chains', which are characterised by mutual interdependence of the parties involved, also tend to decentralise knowledge towards the local level. Partners acting interdependently often need knowledge abroad to solve particular problems. 'Modular value chains' stimulate local knowledge abroad beyond the standardised interfaces, as modularisation does not entirely overcome the need for coordination and collaboration. Knowledge is therefore needed at the local level too. Last not least, 'arm's length market relations' or transfers on anonymous markets also require local knowledge. This is particularly because such relationships are regulated by standards and certificates and thus require organisational learning in the production process (Gereffi, Humphrey and Sturgeon 2005). There are several cases today which show that export production for the global markets, along with the need to meet international quality standards, represents a substantial impetus for organisational learning.

At the same time, export production for the world market cannot be regarded as the only and ideal way towards industrial upgrading in worldwide production plants. Navas-Alemán (2011) analysed a Brazilian furniture and footwear cluster. She reports that domestic and regional value chains sometimes offer better opportunities for upgrading, as they enable relatively high value-added activities, well remunerated jobs and unique products with access to various value chains.

At the same time, *global labour* also plays a role. The 'global race for talent' not only applies to R&D, but also to skilled work in production and administrative tasks. Capable yet low-cost labour is sourced at a global level both in the case of white-collar labour (such as engineers) and blue-collar work. Global companies attempt to establish efficient 'world talent pipelines' by gaining, training and retaining staff on a global level; recent management guidebooks indicate the predominance of this issue in human relations today (Gordon 2009, Lockwood 2010).

The increase in the demand for skilled labour coincides with recent technological and organisational change. The implementation of advanced technologies or new organisational arrangements critically relies on skilled employees. Workers need to show a high degree of responsibility for the products of their work and the expensive machinery they use. Meeting the high quality standards of world market production is only possible with a capable workforce. Yet, appropriately skilled *and* cheap labour is scarce, and the available competencies on the worldwide labour markets are often opaque. Particularly in emerging economies such as China, local managers are finding it difficult to gain and retain an adequately skilled workforce. Although unemployment and underemployment are still very high in such countries, employees with the right set of skills are rare, in particular junior executives. In those suitable candidates recruited by a subsidiary, expectations with respect to income and a rapid career path are often high. Hence, companies frequently train their employees in order to engender loyalty to the company.

It can thus be said that customer requirements stimulate top management to decentralise competencies into the subsidiaries abroad. Shortages of skilled workers on the global labour markets motivate headquarters to initiate vocational and further training at international sites. Thus they gradually establish 'integrated production sites'.

To illustrate the concept and emergence of 'integrated production sites', the following section uses automobile suppliers in Poland as an example. This example incidentally also illustrates the huge diversity of peripheries in the world economy, ranging from the least developed countries of the Global South to 'semi-peripheral' countries in Central Europe (Dománski et al. 2013: 152, 157). Poland can be considered a semi-periphery of the large existing markets areas in Western Europe (Fuchs and Winter 2008).

The example illustrates the power of markets: the requirement to produce to world market standards leads to technical and organisational certification on the shop floor and the introduction of ISO standards as a prerequisite for meeting the requirements of international car manufacturers. At the same time, the example once more illustrates that such change needs time and usually proceeds gradually (Fuchs 2003b, Fuchs and Winter 2008). Once a multinational company has made its first investment in a region with an adequately skilled workforce – in particular workers with experiences in the manufacturing industries – managers tend to further invest in skills. Companies sometimes achieve considerable upgrading involving product and processes innovation and extension and deepening of production, ultimately leading to integrated production plants.

TRW in Poland

TRW Automotive is one of the largest automobile supply companies and a global leader in active and passive safety systems, with headquarters in Livonia, Michigan (USA). Today, TRW employs about 65,000 persons worldwide in approximately 185 facilities in 26 countries, including 13 test tracks and 22 technical centres (TRW 2012, 2013a). In Poland, there are about 6,000 employees working in five plants (Pruszków, Czechowice-Dziedzice, Czestochowa, Gliwice and Bielsko Biala). Upgrading is described for the production plants of Pruszków and Czechowice Dziedzice which were formerly peripheral production sites and have now become subsidiaries with considerably more competencies. This is driven by the customers, which are the large car manufacturers.

In the communist era, *TRW in Pruszków* was a cooperative. Later the firm was owned by Fiat, until TRW took it over in March 1999. At the same time, the company moved from Warsaw to a nearby industrial park in Pruszków. As a former Fiat plant, the company started out as a supplier of keys and locks for the smallest Fiat models. It was merely an assembly plant and extended workbench. But management had the shared vision to diversify the client base and sell locking systems to various car manufacturers. Thus in the early 2000s, local managers saw casting as important for process upgrading in order to serve the changing market requirements, and management purchased a number of casting machines. Workers and technicians were trained to operate the new machines, and TRW headquarters offered technological support. Since the early 2000s, the plant implemented new systems of quality improvement and international certification, particularly ISO standards to achieve international quality standards for products and the production process. Additionally, innovations in steering locks and ignition systems were introduced and some production lines reorganised (TRW 2013b). Thus, the plant changed from an assembly line in times of communism to a Fiat plant after the fall of the Iron Curtain to a supplier of various international automobile companies – and thus an integrated production site.

The sister plant of *TRW in Czechowice Dziedzice* tells a similar story of combined product and process upgrading in response to market requirements. This company also initially belonged to Fiat; and TRW acquired it in 1995. The client base was similarly extended to include other brand name car manufacturers, meaning that market requirements were now those of the automotive world markets. Therefore, challenges had to be overcome. Immediately after buying this brownfield investment, TRW identified significant problems with regard to quality and work discipline, so management developed a shared vision on how to respond to such problems. The strategy began with the replacement of all executives in the first six years; only the production manager and the technical manager stayed in their jobs. The qualification requirements were higher than in the first case study, because the production of steering gears requires a larger share of technicians and engineers. Thus, the employees were trained for six months in the subsidiary in Czechowice Dziedzice and other European TRW plants. The plant also implemented international standards such as ISO. These organisational process innovations went along with product upgrading. Steering gears were improved with regard to their stability and functioning. In addition to the existing Italian production machines, management bought new technologies from firms in Japan and Germany. Furthermore, management reorganised production flow and reduced lead times. After these changes, the plant needed less than a tenth of the time for production. As the level of competencies in the second example was higher than in the first case study, the path towards an 'integrated production site' was broader in the plant in Czechowice Dziedzice.

To conclude, both cases demonstrate the relevance of market-related upgrading in the production process and the development from a simple assembly plant to an 'integrated production site' (Fuchs and Winter 2008, Winter 2008, 2009). Both case studies illustrate the importance of technical and organisational certification on the shop floor. The steps involved in introducing ISO were particularly relevant since achieving such standards is a good precondition for linking up to or staying in the global value chains or markets (Schmitz 2004b, Nadvi 2008). At the same time, both case studies highlight the importance of skilled labour. TRW invested in Poland, which is a country with a skilled and experienced workforce; yet it was necessary to further train the employees.

Honeywell in Mexico

Another example is Honeywell in Mexico, where quality improvements play an important role. This is another case study from the periphery of a large existing market area, highlighting a similar situation in a different regional setting. Emphasis is less on technology than managements' shared visions and patterns of interpretation about the employees' work attitudes.

The example is a Honeywell subsidiary in Northern Mexico, which improved its quality standards by implementing a particular 'local' shared vision. Management had the shared vision that the principle of 'mañana', which they ascribed to the Mexican workers, could be substituted by the concept of the 'here and now'. Managements' shared visions about 'mañana' and 'here and now' relate to patterns of interpretation regarding responsibility at work and the binary codes of 'we' and 'the others'. The case was presented by a former manager of the Honeywell plant. The study was part of a research project on upgrading in the Mexican *maquiladora* (Fuchs 2003a) and updated in 2013.

The Honeywell site is an electronics producer located in the city of Chihuahua which started production in 1976. To improve quality, the plant managers initially attended a training course at a prestigious management consultant in the USA supported by headquarters. However, the managers found the consultants' global principles unconvincing and began to develop their own shared vision and strategy. They asked the local works psychologist for advice. Gradually, they arrived at the shared vision that the difference between the theoretical concepts of the US consultant and social reality in Mexico was that Mexican people would not be orientated in the 'here and now', neither in life nor at work. This, so they assumed, was due to the influence of the Roman Catholic Church and its teachings of the unimportance of the individual. Seeing the individual as 'the light of a small candle in infinite darkness', so they thought, could lead to patterns of interpretation about non-responsibility.

The shared vision on 'here and now' touches upon the responsibility workers feel towards work and life in general. Such common mental constructs may resemble prejudices; at the very least it ascribes certain cultural and religious attributes to 'the other'. As it stood, the shared vision held by management affected the chosen strategy. The local managers concluded they needed to increase the workers' awareness of the 'here and now' both at work and at home. They then set out to emphasise the 'here and now' at work, for example by actively helping some workers at the plant to overcome alcoholism. They told the workers not to relinquish alcohol 'mañana', but to give up drinking today in the 'here and now' (Fuchs 2003a: 116–17).

The strategy appeared to be successful. In the early stages of production, the failure rate was about 14 per cent, but managers were able to reduce this to less than 0.001 per cent. In 1992, the site thus became the first company in the Mexican state Chihuahua to implement the quality standard ISO 9000.

Today, the Honeywell Corporation has implemented modern quality standards and certifications at all global production sites. Honeywell is an important employer not only in Chihuahua but in different parts of Mexico with nearly 15,000 employees at 14 production facilities. Apart from Chihuahua, there are plants located in Baja California, Nuevo Leon, San Luis Potosi and Mexico City, supplemented by the Research and Technology Center in Mexicali (Honeywell 2013).

Managements' shared vision, which essentially linked the principle of 'mañana' to the Mexican work attitude and the 'here and now' to the US work attitude, can be related to patterns of interpretation about responsibility and the binary code of 'we' and 'the other'. 'We' in this case is linked to 'Western thinking', modernity and progress, adaptation to the new requirements and to the 'New Mexico', whilst 'the other' is linked to more 'traditional' values. Such binary distinctions are essentially general patterns of interpretation which are commonly applied to diverse settings; they also persist as socio-cultural rules. It is this permanence which distinguishes patterns of interpretation from shared visions which are related to particular situations (see Chapter 5). In the Honeywell case, managements' shared vision had a situational trigger (quality problems); it also referred to general patterns of interpretation (responsibility and the binary code of 'we' and 'the other'). While the shared vision about 'mañana' and 'here and now' was a dynamic phenomenon and disappeared when the quality problem was solved, the pattern of interpretation about 'we' and 'the other' continues to persist in society.

The cases of TRW and Honeywell already implicitly referred to the particularities of the production process and the requirements of local problem solving that may arise in the foreign subsidiaries of multinational companies. The following section shows that these challenges can promote the development of integrated production plants.

Particularities of the Production Process and Local Problem Solving

Reasons for the internationalisation of production-related knowledge are frequently related to the problems arising from the material and process-related specificities of production (Kern and Schumann 1987). Each production process has its particular technology and local requirements.

Local problem-solving in the production process abroad often requires decentralisation of competencies in production, maintenance and repair. Such local problem solving reduces transaction costs as it enables pragmatic solutions to be found on the shop floor in the subsidiaries. Local problem-solving abroad is often easier than long and difficult interactions between headquarters and the foreign plants. As illustrated in Chapter 11, even the internationalisation of R&D-related activities is indispensable for resolving local problems in R&D and production. The following exemplifies the particularities of the production process and local problem solving resulting in the internationalisation of production-related knowledge.

Kirchhoff Polska in Mielec

The roots of the company can be traced back to 1785, when the Witte Needle Factory started production in one of the early metal-working clusters in present day North Rhine-Westphalia. At the end of the nineteenth century, production expanded into wagon parts and later into the automobile industry. Internationalisation of Kirchhoff Automotive began in the 1980s. Today the company is a group with manufacturing and R&D centres worldwide, employing more than 10,000 persons (Kirchhoff 2013a). Apart from expansion to North America and East Asia, the expansion of locations in Central and South Eastern Europe is particularly important (see Map 12.1).

The subsidiary Kirchhoff Polska, located in Mielec, is a supplier for body structure modules as well as chassis and power train components for car manufacturers. Local management initiated an upgrading process towards an integrated production site by establishing and expanding a tool design unit in the plant in the 2000s. Creating applied engineering activities for local maintenance is efficient because of the particular type of production, which relies on specific tools and pressing. The (rather large) parts for pressing differ from one customer order to the next, and the production of each part requires specific pressing moulds which are usually of significant size and weight. Transporting them to the German parent company and back would prove inefficient, so there was good reason for locating tool design and production at the local plant. Maintenance in this case not only implies the repair of broken parts, but also the design and production of

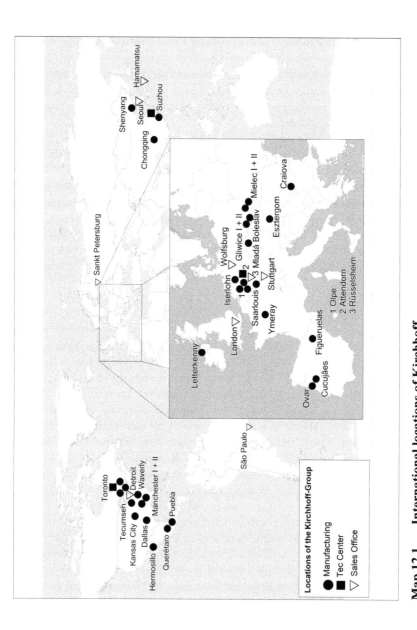

Map 12.1   International locations of Kirchhoff

Source: Based on Kirchhoff 2013a; cartography: J. Mauersberger.

new tools – in other words: process engineering (Fuchs 2005). Practical solutions which improve local competencies are thus strongly intertwined with the material production process.

Upgrading processes such as the one described are a precondition for further upgrading and thus for organisational learning. The local plant gradually improved other competencies and technology. Apart from innovation in material flow and logistics, the site today not only operates and maintains mechanical presses, manual spot welding and standard resistance welding machines, but also welding and fusion welding robots as well as further machines and equipment for various processes. The plant holds a broad range of engineering competencies to keep such technologies running (Kirchhoff 2012: 49).

Upgrading towards an integrated production site can also be traced back to the earlier history of the company, which is a story of continuous learning on the shop floor. The construction of tools is among the traditional core competencies of Kirchhoff and its tool-making sister-company 'Witte Werkzeuge'. The company has always understood tool making as an 'art', considering well-made tools the heart of production in the sense that high quality products can only be made using top quality tools (Kirchhoff 2013b). Since tool-making is deeply rooted in top management's shared vision, the respective tool-making competencies are considered highly relevant for the foreign plants too. In Poland, the local plant manager also actively pushes such processes (Fuchs 2005).

Remarkably, in the academic literature on the globalisation of manufacturing industries and upgrading, such particularities of tangible, substantial production are rarely discussed. On the one hand, regional studies and economic geography tend to focus on the spatial organisation of production at a more abstract level rather than consider actual problems in production. On the other hand, studies in production management and organisation sciences – although they may well acknowledge the complications of production technology – usually fail to take a spatially differentiated perspective. This results in a rather underexplored island in between the popular strands of academic literature (Phelps and Fuller 2000, Phelps and Wood 2006).

In economic geography, such divergent views can be traced back to the early academic discourse. One of the two strands of literature that do refer to the internationalisation of production goes back to the 1980s discussion of the worldwide division of labour and global sourcing. Fröbel, Heinrichs and Kreye (1980) referred to 'The New International Division of Labour' on a macro-level whilst Clarke (1985) studied 'The Spatial Organisation of Multinational Corporations' on a micro-level. Dunning (2000) focused on the link between localisation, internationalisation and intra-firm industrial organisation. Although these studies consider space and organisation to be important, they tend to ignore the detailed material processes of production.

The other stand of literature analyses the tangible production process and local problem solving. This however has the tendency to ignore the difference between parent companies and foreign subsidiaries, focusing instead on companies in the mature core economies. Early stimuli came from empirical studies in the 1970s, which found that technology does not determine the organisation of the production process. The organisation of a Fordist assembly line had long been considered to predefine blue-collar work as low-skilled work, so finding broad elbow room for work organisation and skilled work was an electrifying insight at that time challenging actors in companies as well as academics (von Behr 2012: 36). One of the early publications was Ellegård (1983, 1989) who studied new types of work organisation at Volvo, in particular group work. Similar to Ellegård in Sweden, Kern and Schumann (1987) discussed the 'limits of the division of labour', focusing on 'new production and employment concepts' in the German metalworking industries. Later, *The Machine that Changed the World* became a very popular book (Womack, Jones and Roos 1990), which is a plea for 'lean production' principles in order to increase global competitiveness. The perspective has thus shifted from technological determinism towards elbow room in skill-supporting work organisation; today it is shifting back slightly. There is recognition of a kind of 'object-control' (Rennstam 2012: 1971), acknowledging that technology is not adaptable to each and every requirement but influences how work is organised and knowledge is distributed. At the same time, this acknowledgement of the relevance of technology does not imply a step back into technological determinism.

Given the relevance of process innovation for national competitiveness *and* human labour, organisational patterns of the production process have attracted public funding. In Germany, public funding has supported joint research by social scientists, engineers and practitioners such as managers, works councils and trade unions, in particular during the 1980s and 1990s. This research has focused on technology, organisation and skills and is currently looking at the effects of internationalisation on production and work in Germany (Hirsch-Kreinsen, Lay and Abel 2012).

Recent studies show that managers now acknowledge the requirements of the production process – and particularly process innovation – as important. Som and Jäger (2012) carried out a survey of 1,594 companies in the German manufacturing industries, finding that managers considered process innovation nearly as important as product innovation. The investigation showed that 79 per cent of the companies carry out product innovation. Still, 71 per cent of the companies also focused on new technical processes and 28 per cent on new concepts of organisation (multiple responses allowed). 'Total Production Maintenance' is the most recent core innovative goal (Som and Jäger 2012: 3–5, 8–9). The relevance of process innovation corresponds to general trends highlighted by the OECD (OECD 2009: 29–30).

Similarly, Kuula, Putkiranta and Toivanen (2012: 119) underline that certain problems in the production process play an important role in managements' shared visions. This particularly applies to quality requirements. In a three-phase longitudinal study conducted in Finland in 1993, 2004 and 2010, they examined changes in the shared visions of management with respect to quality. The initial shared vision was 'defects will happen, inspect them out, accept cost of scrap and rework, ship product and deal with customer complaints'. This then changed to 'inspection and control, some data collection to regulate variance, some employee involvement, some rework'. Finally, in 2010, the shared vision had become 'zero defect, total quality mindset, quality controlled in process, quality designed into manufacturability, quality is everyone's job'. But improving quality by changing workers' mindsets is not a new phenomenon. Honeywell was an early example; Fleury and Humphrey (1993) similarly focused on human resources in their study of quality methods in Brazilian manufacturing industries.

Broader still, there are some few studies which relate to the particularities of the international production process. Gutierrez for example (2012: 134, 152–3) emphasises the relevance of process innovation in El Salvador, Guatemala, Honduras and Nicaragua, suggesting that process innovation facilitates product innovation linked to well-developed customer relations. Srinivas and Sutz (2008: 6) focus on the ability to innovate under 'scarcity' conditions in developing countries, describing process innovation as an idiosyncratic way of pursuing appropriate outcomes. Other authors have studied socio-cultural challenges to the implementation of new work organisation (Cagliano et al. 2011) and the influence of socio-cultural contexts on manufacturing strategies (Alas, Kraus and Niglas 2009).

After considering the relevance of product and labour markets and the particularities of the production process for the development of integrated production sites, the following section highlights localised interactions with actor networks and institutions. Particular focus is on vocational education and further training of workers in the subsidiaries of multinational companies in the peripheries of the world economy. The notion of 'integrated production site' is now extended beyond the multinational company to include aspects of regional integration.

Localised Interaction with Actor Networks and Institutions: Vocational Education and Further Training

Local interaction begins early for multinational companies establishing a new production plant abroad. Local communication and networking are required both for the initial entrance into new global markets and the first steps of starting up production abroad. Conditions at the new site may still be opaque at this stage, such as different academic degrees which make hiring a challenge. Early collaboration with local partners is a way of combining global and local knowledge, enabling management to obtain knowledge (Fuchs and Scharmanski 2009).

This implies that foreign direct investment does not transfer knowledge from the parent company to the foreign production plant in a unilinear way. Rather, communication between the global and local partners generates 'glocal' knowledge (Swyngedouw 1997) combining global and local knowledge (Alvstam and Ivarsson 2004). Regional integration of subsidiaries can therefore be considered an instrument for increasing the performance of the entire multinational company (Andersson and Ejermo 2005, Teirlinck and Spithoven 2008).

This relates back to the idea that internationalisation helps to 'tap' into the locations of new and superior technologies (Håkanson 1995: 121–2). However, 'tapping' into foreign pockets of excellence has so far mostly referred to investment in global design centres in parts of New York, London, Paris, Shanghai and Bangalore, for instance for creative work (see Chapter 7) and particularly for R&D (see Chapter 11). Multinational companies connect with other international subsidiaries, local firms, other organisations and institutions through context-specific practices (Manning, Sydow and Windeler 2012: 1202–3). However, they do so not only in specific creative places, but also at the many locations where they invest in production plants.

By entering and then remaining in new investment regions, companies gradually learn how to successfully produce abroad, acquiring more and more knowledge in the process. While the initial contacts to new markets and the production stages are strongly characterised by uncertainty, advances in production and the continuous experiences gained by workers on the shop floor gradually overcome much of this opacity. Over time, day-to-day tasks generate considerable knowledge, and ambiguities are reduced through an ongoing process of learning and responding to different requirements.

This is not only relevant for the production site abroad, but to some extent to the company as a whole. Quoting the well-known saying 'there's nothing like travel to broaden the mind', 'firms learn a lot by going abroad' (Ebersberger and Herstad 2012: 274, 287). Although the foreign subsidiary remains physically distant from the parent company, the organisational distance (Torre and Rallet 2005) and institutional, cognitive and social distance (Boschma 2005) continues to shrink from the point of first market entrance and the initial production stages.

Local integration of multinational companies is more than a particular kind of organisational learning for the company. By 'becoming locals', multinational companies show their willingness to listen to the local markets and clients, their commitment to producing in the particular country and their willingness to empower local staff and invest in their skills and training. Overall, they demonstrate socio-cultural openness, indicating they are not merely imposing global business models and standardised practices on the local settings (McKinsey 2012b). Insofar, local integration also has effects on marketing and placing brands locally.

In many cases, localised interaction of a subsidiary particularly aims to improve vocational education and further training of employees who work in the plant. In this context, the 'integrated production site' is a production plant which is integrated into local and multiscalar networks and institutional arrangements in order to improve the capabilities of the workers. The following example of multinational subsidiaries in Ciudad Juárez illustrates the point.

Interaction with Local and Multiscalar Actors and Institutions: Subsidiaries in Ciudad Juárez

The *maquiladora* industries not only stand for increasing engineering activities and partially elaborated R&D centres in a periphery (see Chapter 11); they also serve to illustrate localised interaction between multinational companies and regional and multiscalar organisations. Even though the *maquiladora* region is rightly characterised as a hub-and-spoke district and satellite platform of production dependent on external headquarters and clients (Markusen 1996), significant local arrangements for vocational education and further training have been set up through networks created by local actors. As a result, the *maquiladora* is no longer a 'tabula rasa', but has skilled workers with experience in the manufacturing industry, thus integrating the manufacturing industries in the region. Even though the *maquiladora* often serves as an example for an archipelago of the world economy – that is part of global economic networks but hardly integrated in the respective region (Veltz 1996) – local subsidiaries are involved in international *as well as* local relations with actor networks and institutional settings.

In Ciudad Juárez the local actors describe their networks and the corresponding institutional settings as a 'triple helix'. This conforms to currently popular approaches which combine governmental support, industrial initiatives and science, with science understood to encompass research and education (Westlund and Li 2013: 131). In Ciudad Juárez, the Universidad Autónoma de Ciudad Juárez provides general education with a broad range of subjects. Furthermore, there are technical universities, such as the public Universidad Tecnológica de Ciudad Juárez and the private Tecnológico de Monterrey, which has its centre in the city of Monterrey and a campus in Ciudad Juárez.

There are various agreements between the universities and local *maquiladora* companies for traineeships and practical training, in cooperation with chambers of industries and the *maquiladora* association. The subsidiaries are active players in this helix in that they not only utilise, but also develop training programmes and organisations together with local actors. Interestingly, the objectives of the training programmes not only comprise technical skills, organisational knowledge, business administration and additional competencies such as English, but also social skills such as the ability to work in a team. Such competencies particularly include experiential knowledge, implicit knowledge and soft skills in manufacturing (Daily, Bishop and Massoud 2012). Apart from the universities, there are further institutions for vocational education and further training which complete the triple helix, such as CONALEP and CENALTEC.

CONALEP

The Colegio Nacional de Educación Profesional Técnica is a central part of practice-oriented education for workers and technicians in Mexico. Initiated in 1979 to improve the qualification standards of the Mexican workforce,

CONALEP has established several locations in Mexico offering different training schemes depending on the regional economic setting. CONALEP is financed and managed by the Mexican federal and state governments and receives support from organisations such as the State Centres for Scientific and Technological Studies and the Institutes of Training for Work.

Ciudad Juárez has a CONALEP site with specific focus on multinational automobile supply companies and the electronic industries. It was established in 1993 with initial World Bank support to install the machines used for training. CONALEP offers technical bachelors' degrees with courses in computer skills and competencies such as turning, drilling, mechanics, hydraulics, mathematical modelling, quality control, implementation of quality standards, and since the 2010s, also skills in mechatronics and industrial electronics. CONALEP closely interacts with local companies, and training is practical, applied and project-oriented. In 2013, students won a prize for designing an anti-theft system and a painting robot (Entre Líneas 2013).

CENALTEC

The Centro de Entrenamiento en Alta Tecnología is another important part of the helix as a provider of locally required industrial skills. The organisation engages local company managers and trainers to continuously adapt its training programmes to technological and organisational change (OECD 2010). Today, CENALTEC offers degrees in metal mechanics, welding, and polymer processing (INADET 2013a). CENALTEC originally began as an initiative of the Philips' machine tool sector which is headquartered in Eindhoven, the Netherlands (Fuchs 2003a). In Ciudad Juárez, a public–private-partnership was created in 2000 together with other local companies and public organisations. The Mexican Ministry of Education gave financial support for buildings, infrastructure and technology; the government of the Mexican state of Chihuahua and other local companies also contributed. Philips' particular contribution was know-how, aiming to implement similar standards in its machine tool making locations worldwide. CENALTEC is the first training centre in Mexico to draw on the Mexican Ministry of Education, the Chihuahua state government and the private sector as three parallel sources of funding (INADET 2013b).

Today, other multinational companies also take part in CENALTEC, including Bosch, Delphi, Epson and Honeywell (INADET 2013b). In 2006 the organisation expanded to the city of Chihuahua where it offers additional training in aerospace and medical device manufacturing. On top of its standardised courses, CENALTEC also provides 'specific accelerated training programmes' with 'tailor-made' schemes to meet the requirements of certain firms. Training courses are flexible with respect to the hours, scope, length and content, and CENALTEC ensures that after completion of the course, trainees have precisely the competencies required by the multinational companies for their production in Northern Mexico. Certifications are endorsed by the American Society of Mechanical Engineers, the National Institute for Metal Working Skills in the USA and the Training Centres for Work of the Ministry of Education in Mexico. Thus, CENALTEC supports the companies in recruiting and training the workforce locally (MFI International 2012).

Training at CENALTEC comprises more than technical competencies and communication skills in English. It also includes quality awareness and particular experiences in manufacturing. Trainers seek to ensure that workers have a thorough understanding of how machines and materials behave, encouraging them to develop a sense for the principles of production. Later such insights are applied to digital processes. Trainers also teach the principles of good planning at work, so that workers understand that precision in work scheduling predefines the quality of the product (Fuchs 2003a: 123–4).

Managements' Shared Visions and Regional Change

Overall, local education and training organisations thus pay attention to experiences, implicit knowledge and soft skills in manufacturing. This corresponds to the shared visions held by the human relations managers, trainers and supervisors interviewed in several subsidiaries in Northern Mexico (Fuchs 2003a: 118–19). Interviews illustrated managers' shared vision that workers' attitude to the quality of production had to be improved so that international standards and certifications could be implemented. Managers were looking for a higher degree of responsibility in their employees (see the Honeywell case above).

Managers also considered training important for another reason: They were keen to tie employees to 'their' company. Loyalty of the workforce is important to the *maquiladora* companies because of the high fluctuation of workers, which generates much friction in the daily organisation of work. Apart from technical training, some local companies also support workers' participation in intense programmes of study to complete their school degrees, such as 'primaria', 'secundaria' or 'preparatoria'. Mexican law favours companies offering such courses to better qualify their employees.

However, it should be noted that such local opportunities for training and industrial restructuring towards upgrading have ambivalent impacts on gender relations (see Box 12.1).

To conclude, the internationalisation of production-related knowledge is promoted by the localised interaction of multinationals with regional actor networks and institutions. Comprehensive education and further training is found in some peripheries of the world economy such as Northern Mexico. These regions advance as nodes in the international networks of knowledge.

The following chapter discusses international labour regulation and labour relations as another potential stimulus for the internationalisation of knowledge and the upgrading of foreign subsidiaries.

Box 12.1   Gender relations in the *maquiladora* region

Upgrading of production processes and the opportunities for vocational education and further training affect female workers in contradictory ways. On the one hand, upgrading implies that increasingly capital intensive technologies are used. Since this requires well-trained employees, it decreases the share of repetitive assembly work, reducing job opportunities for the low-skilled female workers which made up a large share of the workforce in the *maquiladora* industries in the 1980s and 1990s. On the other hand, upgrading and training brings new job opportunities for well-educated women such as female engineers and commercial directors. Although still a long way from gender equality, today's young women in Mexico have much better career opportunities than their mothers and grandmothers had (Catanzarite and Strober 1993, Cravey 1997). Positive examples of successful female friends or relatives encourage young women to imagine the 'American Dream', or at least develop a vision of their own financially independent existence (Fuchs 2001). In other regions of the world, education is often still confined to the sons of the families at the exclusion of daughters (Homm and Bohle 2012: 290–91).

International Labour Regulation and Labour Relations as Potential Stimuli for Upgrading

International labour regulation and labour relations can represent another stimulus for the internationalisation of knowledge and subsidiary upgrading. Multinational companies are shaped by power and the strategies of different parties (Geppert and Dörrenbächer 2011: 6), making them 'contested terrain' in which stakeholders enforce their interests and 'bargain over globalisation'. Labour relations can be 'tool kits' for such socio-political strategies (Williams and Geppert 2011: 74–5).

Consequently, the International Labour Organization (ILO) attempts to exert influence on multinational companies. In the view of the ILO, companies benefit from the implementation of international labour standards, which are usually designed to improve working conditions and ensure that managers respect workers' representations such as trade unions and works councils. Thus, the ILO suggests a 'high-road' path of socio-economic development:

> Higher wage and working time standards and respect for equality can translate into better and more satisfied workers … . Investment in vocational training can result in a better-trained workforce and higher employment levels. Safety standards can reduce costly accidents and health care fees. Employment protection can encourage workers to take risks and to innovate. (ILO 2013)

However, as will be shown below, international labour regulation still has little influence on the internationalisation of production-related knowledge and local upgrading in the peripheries of the world economy – with some noteworthy exceptions.

An important reason for this weakness is that labour regulation is a national matter and thus varies from country to country. In many countries of the Global South, domestic labour relations are shaped by patriarchal relations rather than formal workers' representations (Werner 2012: 412); often, there is no formalised workers' representation at all. As a result, 'evidence of initiatives to build global structures of employee representation remains fairly sparse' (Rüb 2002: 5).

At the *global level*, there are the International Trade Union Confederation (ITUC) and the global trade unions of the respective industrial sectors. These are supported by the ILO which operates within (and influences) the Global Compact of the United Nations. The aim of these organisations is to improve working conditions in production plants worldwide. Some unions have arranged international framework agreements with multinational companies (Franz 2010, Wills 2002). More such international arrangements have been made over the last decades.

In part, the process is also driven by management keen to shape labour relations according to their own corporate standards in the worldwide subsidiaries (Fleming and Jones 2013, Jullien and Pardi 2013: 103). International workers movements are therefore orchestrated by diverse initiatives which are often referred to as 'civil regulation' (Jagodzinski 2012: 21–2).

Corporate Social Responsibility (CSR) has become a popular topic both in the media and in academia. But there is little documentation of the effects of such principles on knowledge internationalisation and subsidiary upgrading (Bergström and Diedrich 2011: 898). CSR partially overlaps with issues of co-determination and participation, which are the traditional remits of trade unions and works councils. In Germany, relevant issues in this context include vocational training, further education, organisational restructuring and plant relocation. If CSR agreements relate to *international* matters, they can indicate fruitful bargaining processes as well as commonly shared visions of managers and international workers' representatives concerning the social standards in the multinational company (Beile, Feuchte and Homann 2010: 8, 17–21, see also Fulton 2007: 9, 56). Thus, CSR agreements on international employment issues represent broadly shared socio-cultural values and standards in the corporation. This is sometimes referred to as 'organisational citizenship' (Wong, Tjosvold and Liu 2009).

Institutional arrangements at the *European level* are more binding than those at the global level, rendering European labour relations much more pronounced and vivid. In 1994 the European Union passed a Directive on the establishment of European Works Councils, for the purpose of informing and consulting workers' representatives in companies with at least 1,000 employees within the EU and at least 150 employees in each of at least two EU member states. From 1994 to 2012, the number of European Works Councils rose from 38 to 1,017 (ETUI 2013a). Although they follow different strategies of participation, they are generally considered effective negotiation bodies (Greifenstein and Kißler 2012: 17, see also ETUI 2013b). Since European Works Councils address issues related to product and process innovation at different locations of multinational companies, they are also concerned with the international division of competencies in the various subsidiaries. The case study of Volkswagen Navarre illustrates how the European Works Council influences the up- and downgrading of production at different sites (Fuchs 2008b).

Box 12.2   International labour relations and upgrading of Volkswagen Navarre

Since the 1990s Volkswagen has considerably expanded its brands and production sites. Today, the corporation has about 100 production sites with about 550,000 employees worldwide, of those about 250,000 are in Germany and 160,000 in other European plants (see Map 12.2). Worker representation has internationalised in line with these developments. The European Works Council at Volkswagen was established in 1990 before the European legislation was passed. In 2009, a 'Charter on Labour Relations' was issued, constituting and setting out the rights of the Group Global Works Council (Volkswagen 2012c).

Volkswagen Navarre started up in the early 1980s when Volkswagen took over Seat in the city of Pamplona (Spain). Since then it has produced the car model Polo as well as engines. But competitive pressure was high. Polos were temporarily also produced in Brussels (Belgium) and Bratislava (Slovakia); in addition to the European locations, there are also international plants for Polo production which serve the global markets (Pune in India, Kaluga in Russia, Pekan in Malaysia, Uitenhage in South Africa and Shanghai in China). On top of this, SEAT and Škoda produce similar models on a common 'platform' in the Volkswagen Group. Although competition between the brands is limited because of different designs profiles, the technology involved in each platform is similar. The platform strategy thus contains some risk of 'cannibalism' between the brands (Freyssenet and Lung 2004: 90).

Given the competitive situation between the international plants, there is an obvious need for cooperation amongst the workers' representatives. German and Spanish workers' representatives began early to coordinate their activities in the European Works Council. Early collaboration was tricky because labour relations at the two sites were different. In Germany, the Volkswagen works council was used to finding cooperative solutions together with management. In Navarre in contrast, competition between different trade unions made it hard for the elected trade union to convince workers that in some economically difficult situations, cuts proposed by management should be accepted. Such bad news was particularly difficult to communicate since the other trade unions offered more popular ideas. Yet over time, the international members of the respective works councils succeeded in finding commonly shared visions and solutions.

In European Works Councils, employee representatives usually have to look for win-win situations with management in order to achieve their goals. Local management in Navarre was interested in keeping and even upgrading production during the recession. In Germany, top management also had good reasons to push the Navarre location, both to demonstrate strong commitment to the country and to be visible with the Volkswagen brand on the Spanish automobile market. There was a shared vision of headquarters in Germany, local management in Navarre and the international workers' representatives to keep and upgrade the location.

Thus recently, Volkswagen Navarre has succeeded in becoming the 'leader production site' for the car model Polo and received priority for the product in Volkswagen's international production network. Additionally, there have been considerable process innovations, particularly in 'greening' the factory and improving environmental standards. Remarkably, during the last years, product innovations have been introduced, too. Still concentrating on the Polo model, the plant has recently received new versions of the model, for example a car with an automatic glass roof, the sporting model (GTI) and the 'green' version of the Polo. Vocational training for junior employees was also augmented (Volkswagen 2013c).

Compared to the global and European level, institutional regulations are even more binding in selected countries at the *national level*. In some countries, law allows for robust co-determination with regard to organisational restructuring, relocation of production, vocational and further training (Scholz 2013). An investigation of German works councils illustrated that a third of those interviewed is involved in innovation processes and actively participate in human relations policy and changes in work organisation and social conditions (Kriegesmann, Kley and Kublik 2010: 72–3).

Additionally, in some European countries large companies have a supervisory board composed of managers and employee representatives. The supervisory board is an institution in addition to executive top management and another source of information for workers' representatives with regard to the international production network. Usually, workers' representatives are in a minority position on the supervisory board, but they are able to influence decisions strategically and tactically (Fulton 2007: 8–9). In Germany, workers' representatives are also involved in the 'Wirtschaftsausschuss' (economic committee), which provides additional information on upcoming strategies. Such activities however are restricted to selected companies in some countries of the core economies.

The above has made clear that the influence of workers' representatives on managements' visions, plans and decisions differs considerably at different spatial levels. The influence of labour relations is relatively strong in corporative frameworks in selected European countries, less strong at the general European level and weak at the global level (Glassner 2012: 89–90) (Table 12.1).

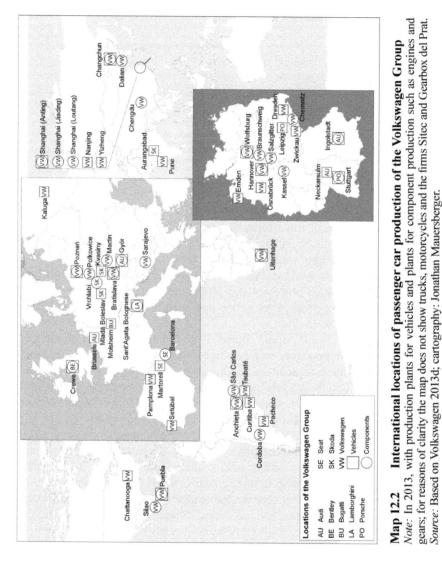

Map 12.2   International locations of passenger car production of the Volkswagen Group

Note: In 2013, with production plants for vehicles and plants for component production such as engines and gears; for reasons of clarity the map does not show trucks, motorcycles and the firms Sitec and Gearbox del Prat.

Source: Based on Volkswagen 2013d; cartography: Jonathan Mauersberger.

Table 12.1   International workers' representation

Level	Company	Sectoral	Trans-Sectoral	
Regional	Local works council respectively unit of the trade union	Regional unit of the trade union	Regional unit of the umbrella organisation of the trade union	I N F
National	Works council, trade union	National trade union	National umbrella organisation of the trade union	L U
European	European Works Council	European trade union	European trade union confederation	E N C
Global	World works council and similar arrangements	Global trade unions	International trade union confederation	E

Source: Based on Rüb, Platzer and Müller 2011: 32, 35, translated and modified.

As shown above, the influence of international labour regulation and labour relations on the internationalisation of production-related knowledge and local upgrading in peripheries of the world economy is generally weak. The few existing exceptions are driven by the shared visions of particular actors engaged in shaping international labour relations and acting in specific institutional arrangements and actor networks.

As shown by the Navarra case study, international participation and co-determination need time to allow international workers' representatives to gradually overcome various difficulties. Apart from differences in labour relations, a common language is often a problem. In formal meetings, communication is facilitated by interpreters. Still, the important tasks of fine-tuning, the exchange of knowledge and sharing visions and interpretations are also important, inside and outside the formal meetings. At the same time, it is precisely this kind of informal exchange which is needed to create organisational citizenship (Wong, Tjosvold and Liu 2009) based on increasingly common shared socio-cultural values and norms (O'Connell, Hickerson and Pillutla 2012).

It should also be acknowledged that works councils and trade unions are elected by the workers of specific locations. They therefore need to represent the immediate concerns of their local voters, such as keeping production capacities high in 'their' plant or receiving new production orders even if local advantages imply a worsening of conditions for the workers elsewhere in the multinational corporation. Thus, the members of works councils and trade unions often feel conflicted between their knowledge of international disparities and ethics of solidarity and their concrete mission of representing the demands of their voters (Cumbers, Nativel and Routledge 2008: 371–3).

Up to this point, issues have been discussed which influence management's shared visions of the decentralisation of production-related knowledge into the subsidiaries in the peripheries of the world economy. Previous sections touched upon the requirements of the clients abroad, access to foreign labour markets, the particularities of the production process and local problem solving abroad, localised interaction with actors and institutions and, last not least, international labour regulation and labour relations. The following discusses the consequences of such processes at the micro- and macro-level. Upgrading particularly affects the company (micro) level, as the following section explains.

Upgrading Towards Integrated Production Sites

Upgrading refers to the increase in competencies of subsidiaries or suppliers in the global value chains. Additional competencies result from product upgrading, process upgrading or adaptation of production to new requirements, such as functional and inter-sectoral upgrading (Gereffi 1999, Gereffi, Humphrey and Sturgeon 2005). A central question is how production sites can move up in the global value chains by maintaining and augmenting their innovativeness and range of tasks.

Since the late 1990s, many authors have suggested that there are – at least some – opportunities for upgrading in international production sites. The topic inspired a number of empirical studies in diverse sectors of manufacturing, such as the apparel industry (Gereffi 1999), the automotive industry (Sturgeon, van Biesebroeck and Gereffi 2008), the agrifood industry (Anim-Somuah et al. 2013) and various other sectors (Schmitz 2004b). Holm, Johanson and Thilenius (1995) detected a trend towards more sophisticated and independent subsidiaries. Subsidiaries were assumed to receive new mandates and tasks (Birkinshaw and Hood 1998, Tavares and Young 2006) and – to some degree – control competencies. Other authors considered subsidiaries to be integrated in rather heterarchical production systems (van Egeraat and Breathnach 2012: 1154–5).

However, such investigations also highlighted the limits of upgrading. Upgrading needs particular institutional settings and specific situations, and results from a window of opportunity which opens up over a particular period of time. Later on, there might be stagnation or downgrading. The trajectories of upgrading are not linear (Nathan and Sarkar 2013: 4), and there is no universal shift of power from the cores to the peripheries. There are diverse ways of upgrading, downgrading and backsliding (Tokatli 2013: 998). A study by Roland Berger (2013a) conducted in 86 companies illustrates that companies indeed tend to combine centralised control with decentralised management, but do so in a limited manner. The study highlights that 60 per cent of the interviewees expect to further 'internationalise their headquarters' in the sense of linking headquarters to international projects. Yet, only 14 per cent expect to relocate headquarter functions abroad; and if so, predominantly into the other parts of the core economies. Only less sophisticated and less sensitive tasks have been relocated to low-cost locations into shared service centres and sites for service outsourcing and offshoring.

The general view on gradual upgrading also has to be qualified with respect to industrial sectors. Those sectors that globalised early are not necessarily those which have undergone considerable upgrading in the peripheries. The garment industry started to globalise as early as the 1970s and 1980s; yet the large 'cut-make-trim' segment of garment production has hardly become knowledge intensive. Some sectors which globalised later started out with a high global distribution of competencies, for example IT software systems design (Bernhardt 2013, Nathan and Sarkar 2013: 5).

Upgrading is also criticised with regard to the problematic relationship between economic and social upgrading. Economic upgrading does not necessarily lead to better working conditions as it is not directly linked to social upgrading, understood as employee skills, incomes, benefits, working conditions, positions and enabling rights, such as the freedom of association and collective bargaining (Plank and Staritz 2013: 4–6, Tokatli 2013: 996). An early study by Kaplinsky et al. (1995: 27–31) emphasises aspects such as skills, incomes and the quality of labour in their study about manufacturing in South Africa. Selwyn (2012a, 2012b) relates chain governance and upgrading to labour relations in Brazil. Recently, Plank and Staritz (2013) have detected 'precarious upgrading' in the electronics industries in Hungary and Romania. International competition puts additional pressure on workers to complete tasks in a shorter time span. This leads to a growing demand for workers to adapt to market requirements and the necessities of the production process. This is flexibility *of* the workers and not *for* the workers in the worldwide plants (Daemmrich and Bredgaard 2013: 167). In this case, increased flexibility means workers' adaptation to production requirements rather than adapting production to individual ideas and concerns. Under such conditions, acquiring new knowledge only partially leads to a broadening of horizons. More often, it is simply an imperative to survive in situations of fierce competition. In many analyses the question of 'upgrading for whom' is still open (Tokatli 2013: 996, 1003–6).

Other studies illustrate a downgrading of plants in the value chains: Bernhardt (2013) detected downgrading with regard to employment and wages during the 2000s in some of the apparel-producing developing countries. Bernhardt and Milberg (2013: 514–15, 525–6) extend this to different economic sectors in developing countries. They find differences in social upgrading between the apparel industries, mobile telecom industries and the tourism sector. They identify some social achievements for the employees in the tourism industry, but hardly any for the apparel and mobile telecom industries.

Sometimes process innovation has negative effects on employment. Whilst process innovation is crucial for plant upgrading and competitiveness on the world markets, it often implies automation, rationalisation, streamlining and the replacement of jobs by technology. Such processes not only occur in the core economies, but also in some subsidiaries of the Global South. This might be surprising since labour is still comparatively cheap and should theoretically not need to be replaced by technology. In the low-cost peripheries, the initial management objective for automatisation is often to increase quality rather than replacing labour costs by capital. If the economic growth effects are not significant enough, such investments in technological equipment, machines and robots will result in job losses.

However, process innovation frequently implies better work, too. Process innovation can improve safety at work and ergonomics. Implemented in the production plants of the core economies, humanisation of work is sometimes 'exported' via new technology to the peripheral plants. Thus, achievements in improvements of work conditions are 'embodied' in technology and transferred from the North to the Global South (Fuchs 2003a, 2008a).

Another important issue needs to be considered. Economic upgrading seems to suggest that production sites develop from low-road and low-tech to high-road and high-tech. Production sites need to modernise and improve production. However, the notion of high-tech is misleading since low-tech production can be important for upgrading towards integrated production plants, too. At present though, low-road development and work experience is hardly discussed in the context of production site upgrading in the peripheries of the world economy. Incidentally, the idea that low-tech could be relevant for the Global South is not new. Similar views existed in the 1960s and 1970s on 'appropriate technologies' or 'intermediate technologies' for the Third World respectively the Global South.

In summary, this section illustrated that upgrading processes towards integrated production sites do exist. At the same time, upgrading is not a unilinear, universal process but selective and related to particular settings and opportunities. Upgrading needs to be differentiated with respect to its economic and social dimensions and qualified with regard to high-road and low-road development. To broaden the perspective, the next section describes indications for the internationalisation of knowledge. Vocational and further training are a special focal point.

Indications for Internationalisation of Production-Related Knowledge into the Peripheries of the World Economy

Apart from upgrading, vocational and further training represent further indications for the internationalisation of production-related knowledge into the peripheries of the world economy. In the context of international production management, training of young employees and the continuous training of older employees is a vibrant topic.

Initiatives of multinational companies to improve the skills of the international workforce often go back to the early expansion of foreign production sites in the second half of the twentieth century. Multinational companies established vocational and further training in their subsidiaries, promoting additional skills by the international exchange of staff.

Today, such skill development is expanded particularly in the emerging markets. The global chemical producer BASF for example created an international training scheme for coating, in which BASF cooperates with vocational schools and coating workshops abroad (BASF 2011). Another example is Bosch, which has established centres for dual occupational training worldwide; recently, such centres have been opened in China, India, Brazil and Vietnam. Training for problem-solving and social skills are particular focal points (Bosch 2013).

The car manufacturers also offer comprehensive programmes for apprenticeship and further training. For example, Toyota and Honda introduced highly specific job training for the workers in their subsidiaries and for the employees of the local subcontractors in the Java automotive cluster (Rutten and Irawati 2013: 146). As Coe et al. (2004: 479) illustrated for BMW in Thailand, governmental and other organisations such as the Board of Investment and the Thailand Automotive Institute worked together to improve the vocational skills of the local workforce. Similar organisations cooperated with other foreign automobile subsidiaries in India. Hence, as set out above for the *maquiladora* region, localised interaction in actor networks and institutions is also important in this context.

Another example is the Volkswagen Group. Recently, Volkswagen has created a global 'Volkswagen Group Academy', which strives to ensure that all employees have the same degree of know-how, expert knowledge and understanding of responsibility for quality. Twenty-six academies have been established in ten countries (2013). Additionally, a 'cultural hub' was implemented at the shop floor level to create more systematic knowledge exchange between the worldwide production plants. Furthermore, mirroring the German system, Volkswagen introduced dual systems of vocational education and further training in its overseas plants based on German standards, organisational patterns and practices. Current examples are the Volkswagen sites in Kaluga (Russia), Pamplona (Spain) and in Chattanooga (USA). In Chattanooga, programmes are certified through the German American Chamber of Commerce in Atlanta, in coordination with the Association of German Chambers of Commerce and Industry (Volkswagen 2012a, 2012b, 2013a, 2013b). Regions thus have different arrangements of public actors, chambers of commerce, trade associations and other initiatives which offer backing for vocational and further training and act as stimuli for local development.

It is noteworthy that some states of the core economies provide active public sector support for domestic company initiatives to increase vocational and further training in their international subsidiaries. Examples include the Federal Institute for Vocational Education and Training (BIBB 2013) in Germany or the Overseas Vocational Training Association (OVTA 2013) in Japan.

Whilst international activities of multinationals to train their workforce are not a new phenomenon, a remarkable and increasingly complex global–local interplay has recently emerged, bringing together multinational companies and public and other organisations for the purpose of improving worker skills.

The degree to which multinational sites engage in improving vocational and further training differs between regions. To some extent, this diversity can be explained by the substantial differences between the various national education and training systems. Additionally, there are no universal training standards for the blue-collar workforce that could guarantee similar skills for multinationals at each global location (Jürgens and Krzywdzinski 2013: 129–30). This last aspect is a challenge for technological as well as other complex skills. As the *maquiladora* case illustrated, companies need employees with competencies in particular technological tools, organisation, English language and additional social skills, including patterns of interpretation such as responsibility for the product and production process. Malish and Vigneswara Ilavarasan (2011: 127–8) describe the ability to work under pressure as another competency required in the Indian software industry; additionally, leadership skills are relevant for executives. In the face of these regional differences, some multinationals have implemented specific skill standards in their different plants worldwide (Schamp and Stamm 2012).

The challenge now is to integrate the insights presented so far in Chapter 12 into general development trends. The next section illustrates that it is difficult to draw a picture of general macro-level trends.

Indications for Change on the Macro-Level

The previous sections have amply illustrated that the international division of production-related knowledge does not simply follow Ricardo's idea of perfect specialisation with highly skilled workers in the North and low-skilled work in the Global South (Gomory and Baumol 2013: 24–5). Micro-level studies have illustrated a broad diversity. So what does the international division of production-related knowledge imply at the macro-level?

Finding answers to this question is tricky. Studies relating to particular sectors of manufacturing point out windows of opportunity for upgrading in the automotive industry (Sturgeon, van Biesebroeck and Gereffi 2008) and various other sectors (Schmitz 2004b). In general, the overall picture is diverse and heterogeneous, especially since the academic literature is mainly based on case studies.

Case studies also influence statements of organisations such as UNCTAD (see UNCTAD 2012: 30, 39–40). From a normative perspective, UNCTAD considers upgrading of production to be relevant for the economy and social improvement in the Global South and other peripheral regions. With regard to the economy, UNCTAD (2013: III) appreciates the growth which results from successful integration into global value chains as long as it is accompanied by upgrading. Still, the limited data available on the macro-level yields little information on upgrading or downgrading on the shop floor in the peripheries of the world economy.

One frequently used indicator is participation in global trade as a way of assessing the degree to which a country's industry is integrated in global production networks. In general, the proportion of global value added trade has been growing in the developing countries during the last decades, rising from about 20 per cent in 1990 and 30 per cent in 2000 to more than 40 per cent in 2010 (UNCTAD 2013:

13). Export and import figures however give little information on innovation, expansion of local tasks or improvement of working conditions. The question remains whether such processes imply substantial upgrading effects on the shop floor and benefits for the employees of the companies in the Global South.

With respect to skills, the OECD (2012b: 408–21) offers data on the participation of adults in education and training. Yet there is no specific information on manufacturing-related skills in the context of foreign direct investments. Figure 12.1 gives an overview of annual labour costs of employer-sponsored non-formal education as a percentage of the gross domestic product. This is training which is planned and organised, but does not lead to certificates. The figure only relates to OECD countries but gives useful information on core-peripheral patterns in Europe, emphasising the high share of Nordic countries, Germany, United Kingdom, Austria, Belgium and the Netherlands and the below average share of countries in Eastern and Southern Europe such as Estonia, the Slovak Republic, Poland, Hungary, Portugal, Spain, Italy and Greece. On average, in the OECD countries about 50 per cent of 25 to 34 year olds take part in formal or informal education, and 27 per cent of the 55 to 64 year olds (OECD 2012b: 408).

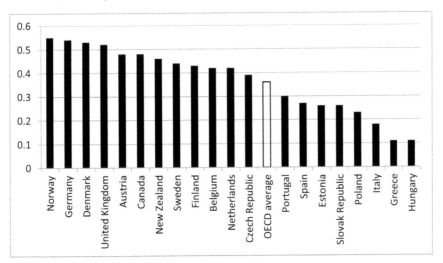

Figure 12.1   Annual labour costs of employer-funded non-formal education as a percentage of the gross domestic product

The OECD (2012c) 'knowledge index' serves as an indicator of the learning environment and the *potential* for knowledge development. Performance scores are used for education and human resources, for the innovation system and for the diffusion of IT in the different countries (Figure 12.2). Again, however, these are no real indicators of knowledge internationalisation and upgrading; instead, they are vague clues that hint at potentials.

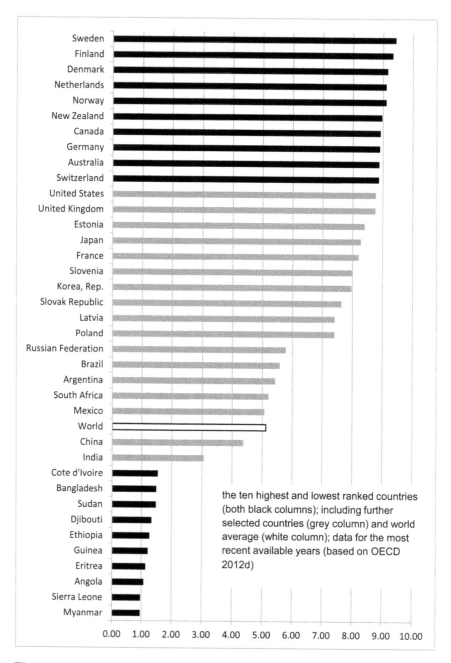

the ten highest and lowest ranked countries (both black columns); including further selected countries (grey column) and world average (white column); data for the most recent available years (based on OECD 2012d)

Figure 12.2   Knowledge economy index

The current index illustrates well-known patterns of the world system. The knowledge economy index of the high income countries is 8.60, that of the upper middle income countries 5.10, that of the lower middle income countries 3.42 and that of the low income countries 1.58 (data for the most recent available years, OECD 2012d). Translated into world regions, the knowledge economy index in North America is 8.80, 7.47 in Europe and Central Asia (which is a rather broad category), 5.32 in East Asia and the Pacific regions, 5.15 in Latin America, 4.74 in the Middle East and North Africa, 2.84 in South Asia and 2.55 in the other African countries. The world average is 5.12 (data for the most recent available years, OECD 2012d). Figure 12.2 shows the knowledge economy index for selected countries.

The available statistical data are therefore clearly insufficient for revealing the effects of the internationalisation of production-related knowledge and upgrading at the country (macro) level. If anything, they suggest very little change at the macro-level in the peripheries of the world system, in particular the countries of the Global South. Thus, the internationalisation of production-related knowledge and upgrading as evident from several case studies is not reflected in major developmental trends. The firm-level findings cannot simply be related to their effects on regions or nations (Tokatli 2013: 993, 1007–8). Studies focusing on the micro-level in multinational companies therefore ought to be complemented by macro-level perspectives (Morgan 2011: 433).

In summary, Chapters 11 and 12 illustrated a general trend towards the further internationalisation of R&D- and production-related knowledge. At the same time, this trend is highly selective with regard to countries and regions, companies and sites, and workers with different skills. A topical question is what impacts this will have on the core economies in the North. The multifaceted and dynamic situation makes it difficult to estimate this. Nevertheless, the following section attempts to sketch out emerging impacts on the North, although this necessarily remains vague and restricted to a few identifiable aspects.

Chapter 13

Impacts on the North: Globalisation of Knowledge as a Race to the Bottom or Industrial Transition with New Arrangements Worldwide

From the expansion of industrialisation in the twentieth century, workers' collective position of power and their individual existence have repeatedly been undermined by plant shutdowns or downsizing due to relocation of production elsewhere. Since the 1960s and 1970s, multinational companies have internationalised considerable parts of labour-intensive production into the subsidiaries in low-cost areas. This trend continued in the 1980s and 1990s when companies increasingly outsourced and offshored further low-skilled activities into independent firms located in the Global South. Substantial relocation of manufacturing resulted, accompanied by significant deindustrialisation in some regions of the core economies. Since the 1990s, the internationalisation of work has been moving up the value chain from assembly up to R&D (Deschryvere and Ali-Yrkkö 2013: 180). After the globalisation of low-skilled work, a worldwide reorganisation of sophisticated tasks is now underway (Horgos 2013: 100), with 'southern engines' (Das and Han 2013: 276–7) increasingly challenging knowledge-intensive production and therefore jobs in regions of the North. Some predict that growing worldwide competition for skilled but low-paid work is likely to cause either a reduction of wages, or growing unemployment of skilled employees, or a combination of both in the countries of the North (Horgos 2013: 100).

Lema, Quadros and Schmitz (2012: 18) argue that by relocating highly skilled tasks globally, the parent companies in the core economies might dig their own grave. Others suggest that relocation could hollow out national innovation systems of the core economies (Deschryvere and Ali-Yrkkö 2013: 180–81). Nathan and Sarkar (2013: 14) see a reversal of the traditional innovation mode that has held fast since the industrial revolution, with relocation of the innovative centres from the mature core economies towards the Global South. Thus, the polarisation between economically strong and innovative regions in the North and the poor regions in the South could possibly reverse.

Such sceptical scenarios are based on the idea of a zero-sum game. This assumes the amount of knowledge to be produced by the world economy to be finite, so the more actors are involved in the game, the smaller the share each player can gain. These static notions clearly ignore win–win situations and dynamics. Hence, Lema, Quadros and Schmitz (2012: 18) also outline another possible future scenario which is based on the co-evolution of the traditional core economies and the newly innovating regions such as Brazil and India. Each region would show economic growth based on different profiles of specialisation.

Although general concerns and sceptical scenarios are much discussed in public life, empirical studies on the impacts on the core economies are actually rare (Deschryvere and Ali-Yrkkö 2013: 181). Selected issues have been studied in various countries, but pronounced processes of relocation in highly skilled work are rarely considered.

With regard to international *R&D* relocation from *Finland to other countries*, Deschryvere and Ali-Yrkkö (2013: 181–2, 200) explain that the general statistical results do not point to a significant short-term relationship between R&D relocation and R&D employment in Finland. However, the authors warn that such results should be treated with care as there are several influencing factors resulting from the particularities of the economic sector and economic dynamics.

With regard to the internationalisation of *services* – which partially also includes the internationalisation of R&D and production-related services – Crinó (2013: 65) analysed data of *nine Western European countries* for the period of 1995 to 2006, revealing weak positive effects on labour demand. For *Germany*, Winkler (2013: 93) found that service offshoring increased in almost all manufacturing industries between 1995 and 2004 and that this slightly reduced the relative demand for white-collar workers in Germany.

For the *manufacturing industries in France*, Besson, Durand and Miroudot (2013: 150–51) note an acceleration of globalisation from 1990 to 2009. The authors detect three types of dynamics. The first is industrial decline as some French industries have no competitive advantage on the world markets and are forced to face up to job losses. Second, there are industries which use internationalisation for growth, and which – in contrast to the first group – expand their workforce domestically and internationally. Third is an in-between group which struggles to survive, using internationalisation for growth as well as improving efficiency gains. The third group encompasses most French industries.

With regard to *manufacturing* in *Germany*, investigations of Fraunhofer ISI reveal the well-known trend of relocating production to the emerging countries, but also a concurrent trend towards 'backshoring', that is bringing back manufacturing that had been relocated to the peripheries of the world economy to the core economies. In the aftermath of the 2008 crisis the rate of companies that newly internationalise production at foreign plants has declined, stabilising at about 8 per cent. Over the same period, the level of backshoring activities has remained stable at about 2 per cent (Kinkel 2012: 696, Zanker, Kinkel and Maloca 2013). Nevertheless, the general level of foreign manufacturing remains high in German industries, with the majority of large corporations and nearly half of the medium-sized companies operating foreign production plants. This internationalisation of work tasks does not necessarily imply a relocation of jobs from the North to the South or other negative effects on incomes and innovativeness in Germany.

Increasingly, internationalisation into new markets seems to generate growth and thus positive effects for the parent companies in the core economies (Zanker, Kinkel and Maloca 2013).

The above corresponds to the study of Tomiura, Ito and Wakasugi (2013: 246–7) for *manufacturing in Japan*, who find that companies that have internationalised their production tend to be more productive than their domestic counterparts.

The different countries and sectors involved in the (few) existing studies make it difficult to detect clear and unambiguous or alarming trends (Horgos 2013). Still, there are certain indirect effects of internationalisation and worldwide competition looming on the horizon for the core economies.

One is the tendency to increasingly use temporary employees in R&D. This brings down labour costs for engineers and technicians in R&D and increases the flexibility of the home base. Already in the 1990s, Michie and Sheehan (1999) identified a growing share of engineers employed by temping agencies in the *United Kingdom*. In *Germany* companies also tend to employ an increasing share of temporary workers in R&D. Young engineers, experts for information and communication technologies and technicians just entering the job market are often limited to temporary jobs initially and forced to wait for a stable work contract with better employment conditions (Will-Zochol 2011: 243–4). Thus, the global race for talent also implies pressure on employment conditions for highly skilled employees in – at least some – R&D departments in the core economies.

A completely different effect resulting from the 'global race for talent' is that companies invest in vocational education and further training in the core economies. In the *Netherlands*, where regional labour markets do not offer enough skilled workers and companies face bottlenecks with respect to staffing, companies increasingly invest in vocational training and further training (van Dijk and Bosch 2003). There are many similar indications for such trends in other countries (Pilz 2012). Relocating highly skilled work tasks abroad in order to tap into the labour markets of the emerging countries is therefore only one option. The 'global race for talent' does not necessarily imply a shift to offshore labour markets; it can also affect company training schemes at home. In Germany, co-determination tends to support such improvements in vocational education and further training (Stegmaier 2012).

This last aspect points to the relevance of regulation and corporate governance in the North. Earlier chapters have argued that organisational learning needs to be explained within each respective context of actor networks and institutional arrangements. This leads to two further lines of thought.

First, collective actors may anticipate possible scenarios and actively initiate institutional change. Rather than representing rigid frameworks, dense actor networks and institutional settings can be adjusted to expected challenges. One example is corporate governance which has adapted to the growing worldwide competition for skilled workers, safe jobs and working conditions in the North, including new arrangements of knowledge-related work in companies of the Global North (Lippert and Jürgens 2012: 40–43). The perspective of anticipating scenarios is particularly stimulating in the light of 'green' innovation and the move away from the carbon economy towards a low-fossil-fuel economy and the use of renewables (Cooke, Parrilli and Curbelo 2012: 21, Hayter and Le Heron 2002: 13–15, Healy and Morgan 2012, Truffer and Coenen 2012). Corporate governance plays a role in locational change via eco-innovation. For example, works councils in Germany frequently play an active part in process innovation, and even sometimes in 'green' product innovation (Molitor 2013a, 2013b).

Second, in their focus on the 'global race for talent', the media tend to refer to high-tech only. But, as repeatedly shown above, radical innovation, top-level scientific research and trade show novelties are not the only forms of important knowledge (Cooke et al. 2011: 6, 11). Asheim (2012: 995) stresses that 'a region's knowledge base is larger than its science base', while Wyckoff (2013: 308) states that 'innovation is more than R&D'. Low-tech is clearly also relevant. Sectors with low R&D intensity have remained stable in large countries of the OECD since the 1970s. In the USA, Japan and many countries of the European Union low-tech sectors generate more than 40 per cent of the industrial value added (OECD 2011). By no means are low-tech industries dinosaurs which only survive in selected market niches (Kirner, Som and Jäger 2009: 5, 83–4, Som 2012). On the contrary, low-tech companies often perform 'intelligent production of simple products' (Hirsch-Kreinsen 2012: 132). Schamp (2012) illustrates a regional example of a traditional industry in the German city of Pirmasens, where an international shoe competence centre was established as a 'knowledge-broker' (Kauffeld-Monz and Fritsch 2010). This promoted successful innovation and organisational restructuring in the region (Schamp 2012: 95, 98). This example highlights successful innovation in 'old' industries in the core economies and illustrates that change does not occur automatically, but is driven by the regulative arrangement in the region composed of joint local initiatives of different public and private actors and associations.

In conclusion, there is little empirical evidence for alarming trends in the Global North resulting from the 'global race for talent'. Conditions remain difficult for international investment in many peripheries of the world economy, and dense actor networks and institutional arrangements in the core economies adapt to anticipated change.

Chapter 14
Summary and Discussion

Space is the 'fundamental stuff of geography' (Thrift 2009: 85). The spatial view on worldwide knowledge gives new insights into global companies and the challenges resulting from the internationalisation of knowledge for local labour and the regions. At the same time, the spatial view reveals gaps in academic knowledge on how managers perceive the globalisation of knowledge, and how globalisation affects particularly blue-collar workers in the Global South. Equally, there are few studies concerned with the effects on the North.

The book discussed three fundamental questions. The first question was:

- What are the conceptual implications of a comprehensive view of knowledge and interpretation, assuming that both direct managements' decisions with respect to the globalisation of knowledge?

Initial conceptual deliberations on knowledge built on recent insights in economic geography, social sciences, international management studies and organisational studies. The book addressed *economic* knowledge, which the academic debate defines as commodity, resource and capital. Scientific-technical knowledge is an essential type of economically relevant knowledge and considered a key asset for the knowledge-based society. Still, it is only a particular dimension of economic knowledge. To broaden and specify the perspective, the 'star of knowledge and interpretation' mapped different dimensions of knowledge, highlighting the importance of scientific-technical knowledge but also other forms of knowledge and interpretation. The star showed that knowledge can be practice or pattern, exclusive or shared, and 'subjective' or 'objectified'.

Using the 'social construction of reality' (Berger and Luckmann (1966/1991) as a conceptual framework not only helps to go beyond scientific-technical knowledge, but opens up a perspective 'beyond knowledge' itself. Here, the concepts of (situational) shared visions and (wide-ranging) patterns of interpretation are productive theoretical tools.

Several examples have highlighted the relevance of managements' shared visions for the internationalisation of knowledge. They guide the globalisation of scientific-technical and R&D-related knowledge as well as production-related knowledge. But shared visions are not restricted to management. Shared visions held by engineers and technicians influence the successful internationalisation of knowledge. International workers' representatives share particular visions, adjust them in response to managements' shared visions, and hence influence corporate decision-making. Thus, the internationalisation of knowledge is not driven by 'factors', but results from different forms of interpretation.

Situational shared visions are complemented by patterns of interpretation, which are widespread and persistent structures or socio-cultural rules. Like shared visions, they also play a role in the internationalisation of knowledge. One example for a pattern of interpretation is trust, for instance trust in (international) partners. Another pattern of interpretation is responsibility, for example the workforce's sense of responsibility for the production process and product quality. A further pattern of interpretation emerges from the interpretive distinction between 'we' and 'the other'. Distance and proximity are also patterns of interpretation. Proximity emerges as a result of the decreasing organisational, cognitive and social distance of international subsidiaries over time.

Compared to shared visions, patterns of interpretation are still a marginal and largely unexplored issue in the academic debate on the internationalisation of knowledge. Their influence on the internationalisation of knowledge thus needs further investigation.

The question how managers and other actors 'subjectively' perceive the globalised world leads us to the second question, which is concerned with 'objectified' globalisation of knowledge.

• What are the implications of the internationalisation of scientific-technical knowledge, particularly R&D? Looking beyond R&D, how does production-related knowledge internationalise?

From the author's own case studies and further academic literature, it has become apparent that many multinational companies have learnt how to globalise knowledge in the last decades. Such learning needs time. Before truly globalising knowledge-intensive tasks, headquarters usually begin with internationalising some low-skill assembly functions into the peripheries of the world economy. Complex production tasks are only globalised if this proves successful. Later, management sometimes also begins to internationalise R&D.

At the same time, this path is not a must, as is evident from those companies that stagnate as assembly plants. Some sites begin as R&D centres, thus starting out and staying on the high-road. Others 'leapfrog' from a low-road to a high-road strategy; and still others must accept downgrading.

The question remains whether R&D supports the performance of a firm or whether a high business performance favours more R&D expenditures. All in all, there is no simple success story of automatic subsidiary upgrading via globalisation and modernisation. Neither is there a story of permanent dependency and inevitable exploitation of the Global South. The internationalisation of knowledge is not a smooth and unbroken process. A window of opportunity might open up for a particular time and then close again. Trajectories have junctions, and one stage of organisational learning does not necessarily lead on to the next. Each step of knowledge internationalisation is an open-ended process. In view of these complex processes inside of multinational companies and their subsidiaries, and thus in spatial interrelations worldwide, the third question addressed potential regional effects:

- What are the regional implications of labour dynamics in the Global South, and what do these same dynamics spell for the North?

Regarding international subsidiaries as dynamic links between headquarters and regional actor networks and institutional settings seems to be a promising approach to better understand multinational companies and their strategies of knowledge relocation (Geppert and Dörrenbächer 2011, Williams and Geppert 2011). Localised interaction between multinational companies and regional and multiscalar actor networks and institutions is very important both for R&D and production since there is 'the firm in the region and the region in the firm' (Schoenberger 1999: 205). Even if there is little scope for 'hard' policy intervention to direct multinational investments, the multiscalar actor networks and institutions influence the fiscal environment, educational facilities and creative atmosphere in the region.

As a result, some regions have gained strong surplus meaning and are considered 'creative' or 'innovative' places, in particular areas in global cities. Apart from these well-known hotspots of knowledge, the peripheries of the world economy hold other and sometimes hidden places for R&D and production-related knowledge. These locations of engineering activities and production upgrading are largely ignored in the academic debate on creativity and knowledge-intensive places. Places such as Ciudad Juárez and other *maquiladora* towns are not ranked highly enough in the hierarchy of global cities to be considered, but they play an important role in their respective organisational–spatial context. Hence, the meaning of 'periphery' is changing in that periphery no longer equals 'marginal'. On the contrary, some peripheries have particular competencies. Still, we know little about the skills of blue-collar workers and about vocational education and further training in regions of the peripheries of the world economy (Pilz 2012).

More sophisticated R&D and the upgrading of subsidiaries in peripheral regions does not mean that these regions will necessarily become knowledge-based core regions. Although the 'global knowledge society' is widely debated, there is no empirical evidence for this kind of 'one world'. Obviously, the dynamics of places attracting more R&D or sophisticated production (particularly in the big emerging markets and peripheries of large existing market areas) cannot change the general socio-economic situation of cities, regions or countries at large. Although innovation in companies sometimes leads to territorial upgrading (Jessop et al. 2013: 116), the increasing knowledge in some subsidiaries in the peripheries of the world economy alone is not enough to overcome poverty in other parts of the population. This becomes even more obvious in view of the huge disparities between urban and rural areas in the Global South (Westlund and Kobayashi 2013).

With regard to the North, at least today there is little empirical evidence for alarming trends resulting from knowledge internationalisation and the 'global race for talent'. In many core economies dense actor networks and institutional arrangements adapt to anticipated change and promote new industrial leadership. The globalisation of knowledge appears not directly comparable to the kind of globalisation of production which focuses on cheap labour worldwide and which often goes hand in hand with a massive relocation of jobs from the North to the Global South. In any case, apart from the globalisation driven by labour-costs, the globalisation of production such as access to new sales markets usually leads to growth effects for headquarters and the parent plants in the North.

Overall, further conceptual and empirical evidence is needed to adequately describe the globalisation of knowledge and the development of skills in the context of regions (Boschma, Iammarino and Steinmueller 2013: 1615). This is not just an academic issue, but also matters to society in that it challenges policy, be it industrial policy, development policy or regional policy (Davoudi 2012). Rather than a simple fact, 'worldwide knowledge' today remains an open question. The conclusion therefore returns to the title of the book: 'Worldwide Knowledge?'

References

Adam, F. and Westlund, H. 2013. Introduction: The meaning and importance of socio-cultural context for innovation performance, in *Innovation in Socio-Cultural Context*, edited by F. Adam and H. Westlund. New York and London: Routledge, 1–21.

Adolf, M., Mast, J.L. and Stehr, N. 2013. Culture and cognition: The foundations of innovation in modern societies, in *Innovation in Socio-Cultural Context*, edited by F. Adam and H. Westlund. New York and London: Routledge, 25–39.

Aglietta, M. 1976. *Régulation et Crises du Capitalisme: L'expérience des Etats Unis.* Paris: Calmann-Lévy.

Agnew, J.A. 2013. Arguing with regions. *Regional Studies* 47(1), 6–17.

Alas, R., Kraus, A. and Niglas, K. 2009. Manufacturing strategies and choices in cultural contexts. *Journal of Business Economics and Management* 10(4), 279–89.

Alexander, J. and Smith, P. 2006. A strong program in cultural theory, in *Elements of a Structural Hermeneutics: Handbook of Sociological Theory*, edited by J.H. Turner. New York: Springer, 135–50.

Alnuaimi, T., Singh, J. and George, G. 2012. Not with my own: Long-term effects of cross-country collaboration on subsidiary innovation in emerging economies versus advanced economies. *Journal of Economic Geography* 5(12), 943–68.

Alvstam, C.-G. and Ivarsson, I. 2004. International technology transfer to local suppliers by Volvo trucks in India. *Tijdschrift voor Economische en Sociale Geografie* 95(1), 27–43.

Alvstam, C.-G. and Ivarsson, I. 2014. The 'hybrid' emerging market multinational enterprise: The ownership transfer of Volvo Cars to China, in *Asian Inward and Outward FDI: New Challenges in the Global Economy*, edited by C.-G. Alvstam, H. Dolles and P. Ström. Basingstoke: Palgrave Macmillan, 217–42.

Amin, A. and Cohendet, P. 2004. *Architectures of Knowledge: Firms, Capabilities, and Communities.* Oxford: Oxford University Press.

Andersson, M. and Ejermo, O. 2005. How does the accessibility to knowledge sources affect the innovativeness of corporations? Evidence from Sweden. *Annals of Regional Science* 39(4), 741–65.

Anim-Somuah, H., Henson, S., Humphrey, J. and Robinson, E. 2013. *Strengthening Agri-food Value Chains for Nutrition: Mapping Value Chains for Nutrient-dense Foods in Ghana.* [Online]. Available at: http://www.ids.ac.uk/files/dm file/ER2BFinal.pdf [accessed: 9 September 2013].

APEX 2013. *Brazil in the World: Innovative, Sustainable, Competitive.* [Online]. Available at: http://www2.apexbrasil.com.br/en [accessed: 20 August 2013].

Arena, R., Festré, A. and Lazaric, N. 2012. Introduction, in *Handbook of Knowledge and Economics*, edited by R. Arena, A. Festré and L. Lazaric. Cheltenham and Northampton: Edward Elgar, 1–20.

Argyris, C. and Schön, D. 1978. *Organisational Learning: A Theory of Action Perspective.* Reading: Addison Wesley.

Arundel, A., Lorenz, E., Lundvall, B.-Å. and Valeyre, A. 2007. How Europe's economies learn: A comparison of work organization and innovation modes for the EU-15. *Industrial and Corporate Change* 16(6), 1175–210.

Arundel, A., Bordoy, C. and Kanerva, M. 2008. *Neglected Innovators: How do Innovative Firms that do not Perform R&D Innovate? Results of an Analysis of the Innobarometer 2007 survey No. 215.* Available at: http://arno.unimaas.nl/show.cgi?fid=15406 [accessed: 30 August 2013].

Asheim, B. and Isaksen, A. 2002. Regional innovation systems: The integration of local 'sticky' and global 'ubiquitous' knowledge. *The Journal of Technology Transfer* 27(1), 77–86.

Asheim, B. and Coenen, L. 2007. Face-to-face, buzz, and knowledge bases: Sociospatial implications for learning, innovation, and innovation policy. *Environment and Planning C* 25(5), 655–70.

Asheim, B. 2012. The changing role of learning regions in the globalizing knowledge economy: A theoretical re-examination. *Regional Studies* 46(8), 993–1004.

Autio, E., Kanninen, S. and Gustafsson, R. 2008. First- and second-order additionality and learning outcomes in collaborative R&D programs. *Research Policy* 37(1), 59–76.

Barney, J.B. 2001. Is the resource-based theory a useful perspective for strategic management research? Yes. *Academy of Management Review* 26(1), 41–56.

Bartlett, C.A. and Ghoshal, S. 1986. Tap your subsidiaries for global reach. *Harvard Business Review* 64(6), 87–94.

BASF 2011. *Globale Lacklehrlinge. [Global Trainees in Painting].* [Online]. Available at: http://www.basf-coatings.com/global/ecweb/de/function/conver sions:/publish/content/press/coatings-partner-magazine/pdf/CP-2011/24-25_ Stampp.pdf [accessed: 30 May 2013].

Bathelt, H., Malmberg, A. and Maskell, P. 2004. Clusters and knowledge: Local buzz, global pipelines and the process of knowledge creation. *Progress in Human Geography* 28(1), 31–56.

Bathelt, H. and Glückler, J. 2005. Resources in economic geography: From substantive concepts towards a relational perspective. *Environment and Planning A* 37(9), 1545–63.

Bathelt, H. and Glückler, J. 2011. *The Relational Economy: Geographies of Knowing and Learning.* Oxford: Oxford University Press.

Bathelt, H. and Li, P.-F. 2014: Global cluster networks: Foreign direct investment flows from Canada and China. *Journal of Economic Geography* 14(1), 45–71.

Beaverstock, J.V., Faulconbridge, J.R. and Hall, S. 2012. Executive search, in *The Wiley-Blackwell Encyclopedia of Globalization*, edited by G. Ritzer. Chichester: Wiley-Blackwell, 615–21.

Beaverstock, J.V. and Hall, S. 2012. Competing for talent: Global mobility, immigration and the City of London's labour market. *Cambridge Journal of Regions, Economy and Society* 5(2), 271–87.

Beile, J., Feuchte, B., and Homann, B. 2010. *Corporate Social Responsibility (CSR) und Mitbestimmung. [Corporate Social Responsibility (CSR) and Co-determination]*. Düsseldorf: Hans Böckler Foundation.

Bell, D. 1973. *The Coming of Post-Industrial Society*. New York: Basic Books.

Belussi, F. and Sedita, S.R. 2012. Industrial districts as open learning systems: Combining the emergent and deliberate knowledge structures. *Regional Studies* 46(2), 165–84.

Benneworth, P. and Rutten, R. 2013. 'Individuals' networks and regional renewal: A case study of social dynamics and innovation in Twente, in *Innovation in Socio-Cultural Context*, edited by F. Adam and H. Westlund. New York and London: Routledge, 185–209.

Berger, P.L. and Luckmann, T. 1966/1991. *The Social Construction of Reality: A Treatise in the Sociology of Knowledge*. London: Penguin.

Bergström, O. and Diedrich A. 2011. Exercising social responsibility in downsizing: Enrolling and mobilizing actors at a Swedish high-tech company. *Organization Studies* 32(7), 897–919.

Berndt, C. 2013. Assembling market b/orders: violence, dispossession, and economic development in Ciudad Juárez, Mexico. *Environment and Planning A* 45(11), 2646–62.

Bernhardt, T. 2013. *Developing Countries in the Global Apparel Value Chain: A Tale of Upgrading and Downgrading Experiences*. Manchester: The University of Manchester, Working Papers Capturing the Gains 31. [Online]. Available at: www.capturingthegains.org/publications/workingpapers/wp_201322.htm [accessed: 21 July 2013].

Bernhardt, T. and Milberg, W. 2013. Does industrial upgrading generate employment and wage gains? in *The Oxford Handbook of Offshoring and Global Employment*, edited by A. Bardhan, D.M. Jaffee and C.A. Kroll. New York: Oxford University Press, 490–533.

Bessant, J., Alexander, A., Tsekouras, G., Rush, H. and Lamming, R. 2012. Developing innovation capability through learning networks. *Journal of Economic Geography* 5(12), 1087–112.

Besson, F., Durand, C. and Miroudot, S. 2013. How much does offshoring matter? Evolution of imports and their relation to profits, labor, and firms' strategies in France, 1990–2009, in *The Oxford Handbook of Offshoring and Global Employment*, edited by A. Bardhan, D.M. Jaffee and C.A. Kroll. New York: Oxford University Press, 123–55.

BIBB 2013. *German Vocational Education and Training Abroad*. [Online]. Available at: http://www.bibb.de/en/wlk15820.htm [accessed: 28 August 2013].

Birkinshaw, J. and Hood, N. 1998. Multinational subsidiary evolution: Capability and charter change in foreign-owned subsidiary companies. *Academy of Management Review* (23)4, 773–95.

Blazejewski, S. and Becker-Ritterspach, F. 2011. Conflict in headquarters-subsidiary relations: A critical literature review and new directions, in *Politics and Power in the Multinational Corporation: The Role of Institutions, Interests, and Identities*, edited by C. Dörrenbächer and M. Geppert. Cambridge and New York: Cambridge University Press, 139–90.

Blomqvist, K., Hara, V., Koivuniemi, J. and Äijö, T. 2004. Towards networked R&D management: The R&D approach of Sonera Corporation as an example. *R&D Management* 34(5), 591–603.

Bosch 2013. *100 Years of Apprentice Workshops: Bosch to Offer Dual Occupational Training Abroad.* [Online]. Available at: http://www.bosch-presse.de/presse forum/details.htm?txtID=6193&locale=en [accessed: 28 August 2013].

Boschma, R. 2005. Proximity and innovation: A critical assessment. *Regional Studies* 39(1), 61–74.

Boschma, R. and Martin, R. 2010. The aims and scope of evolutionary economic geography, in *The Handbook of Evolutionary Economic Geography*, edited by R. Boschma and R. Martin. Cheltenham and Northampton: Edward Elgar, 3–39.

Boschma, R., Iammarino, S. and Steinmueller, E. 2013. Editorial: Geography, skills and technological change. *Regional Studies* 47(10), 1615–17.

Boudreau, J.-A. 2007. Making new political spaces: Mobilizing spatial imaginaries, instrumentalizing spatial practices, and strategically using spatial tools. *Environment and Plannung A* 39(11), 2593–611.

Boulding, K.E. 1956. *The Image: Knowledge in Life and Society.* Ann Arbor: University of Michigan Press.

Bourdieu, P. 1986. The forms of capital, in *Handbook of Theory and Research for the Sociology of Education*, edited by J.G. Richardson. New York, Westport and London: Greenwood Press, 241–58.

Boyer, R. 1987. *La Théorie de la Régulation: Une Analyse Critique.* Paris: La Découverte.

Boyer, R. 1995. Aux origenes de la théorie de la régulation, in *Théorie de la Régulation: L'état des Savoirs*, edited by R. Boyer and Y. Saillard. Paris: La Découverte, 21–30.

Cagliano, R., Caniato, F., Golini, R., Longoni, A. and Micelotta, E. 2011. The impact of country culture on the adoption of new forms of work organization. *International Journal of Operations and Production Management* 31(3), 297–323.

Calantone, R.J., Cavusgil, S.T. and Zhao, Y. 2002. Learning orientation, firm innovation capability, and firm performance. *Industrial Marketing Management* 31(6), 515–24.

Cannon-Bowers, J. and Salas, E. 2001. Reflections on shared cognition. *Journal of Organizational Behavior* 22(2), 195–202.

Cantwell, J., Dunning, J.H. and Lundan, S.M. 2010. An evolutionary approach to understanding international business activity: The co-evolution of MNEs and the institutional environment. *Journal of International Business Studies* 41(4), 567–86.

Capello, R. and Dentinho, T.P. 2012. Globalization trends and their challenges for regional development, in *Globalization Trends and Regional Development. Dynamics of FDI and Human Capital Flows*, edited by R. Capello and T.P. Dentinho. Cheltenham and Northampton: Edward Elgar, 1–12.

Carmona, S. and Grönlund, A. 1998. Learning from forgetting: An experiential study of two European car manufacturers. *Management Learning* 29(1), 21–38.

Carrillo, J. and Hualde, A. 1999. Maquiladoras en redes: El Caso de Delphi-General Motors [Maquiladoras in networks: The case of Delphi-General Motors], in *Enfrentando el Cambio [In Face of Change]*, edited by H.J. Nunez and S. Babson. Puebla: Benemérita Universidad Autónoma de Puebla, 369–85.

Carrillo, J. 2013. *Evolution of the Auto Industry in Mexico*. Paris: Journée du Gerpisa. [Online]. Available at: http://www.colef.mx/ApWp-JCarrillo/ [accessed: 31 March 2013].

Castells, M. 2000. *The Rise of The Network Society. The Information Age: Economy, Society and Culture. Vol. 1*. Chichester: Wiley.

Catanzarite, L.M. and Strober, M.H. 1993. The gender recomposition of the maquiladora workforce in Ciudad Juárez. *Industrial Relations* 32(1), 133–47.

Chen, M.-H., Chang, Y.-C. and Hung, S.-C. 2008. Social capital and creativity in R&D project teams. *R&D Management* 38(1), 21–34.

Chesbrough, H.W. 2003. *Open Innovation: The New Imperative for Creating and Profiting from Technology*. Boston: Harvard Business School Press.

Clarke, I.M. 1985. *The Spatial Organisation of Multinational Corporations*. London: Croom Helm.

Coe, N.M., Hess, M., Yeung, H.W., Dicken, P. and Henderson, J. 2004. 'Globalizing' regional development: A global production networks perspective. *Transactions of the Institute of British Geographers* 29(4), 468–84.

Coe, N.M., Dicken, P. and Hess, M. 2008. Global production networks: Realizing the potential. *Journal of Economic Geography* 8(3), 271–95.

Continental Powertrain 2012. *Facts and Figures 2012*. [Online]. Available at: http://www.conti-online.com/generator/www/de/en/continental/automotive/general/download/daten_fakten_pt_en.pdf [accessed: 26 February 2013].

Cooke, P., Heidenreich, M. and Braczyk, H.J. (eds) 2004. *Regional Innovation Systems*. London: Routledge.

Cooke, P. 2009. Knowledgeable regions, Jacobian clusters and green innovation, in *Knowledge in the Development of Economies: Institutional Choices under Globalisation*, edited by S. Sacchetti and R. Sugden. Cheltenham and Northampton: Edward Elgar, 67–90.

Cooke, P., De Laurentis, C., MacNeill, S. and Collinge, C. (eds) 2010. *Platforms of Innovation: Dynamics of New Industrial Knowledge Flows*. Cheltenham and Northampton: Edward Elgar.

Cooke, P., Asheim, B., Boschma, R., Martin, R., Schwartz, D. and Tödtling, F. 2011. Introduction to the handbook of regional innovation and growth, in *Handbook of Regional Innovation and Growth*, edited by P. Cooke, B. Asheim, R. Boschma, R. Martin, D. Schwartz and F. Tödtling. Cheltenham and Northampton: Edward Elgar, 1–23.

Cooke, P., Parrilli, M.D. and Curbelo, J.L. 2012. Introduction, in *Innovation, Global Change and Territorial Resilience*, edited by P. Cooke, M.D. Parrilli and J.L. Curbelo. Cheltenham and Northampton: Edward Elgar, 1–21.

Cravey, A.J. 1997. The politics of reproduction. Households in the Mexican industrial transition. *Economic Geography* 73(2), 166–86.

Crawshaw, R. 2013. Guest editor's introduction: Politics, economics and perception in regional construction. *Regional Studies* 47(8), 1177–80.

Crinó, R. 2013. Service offshoring and labour demand in Europe, in *The Oxford Handbook of Offshoring and Global Employment*, edited by A. Bardhan, D.M. Jaffee and C.A. Kroll. New York: Oxford University Press, 41–71.

Cumbers, A., Nativel, C. and Routledge, P. 2008. Labour agency and union positionalities in global production networks. *Journal of Economic Geography* 8(3), 369–87.

D'Agostino, L.M., Laursen, K. and Santangelo, G.D. 2013. The impact of R&D offshoring on the home knowledge production of OECD investing regions. *Journal of Economic Geography* 13(1), 145–75.

Daemmrich, A. and Bredgaard, T. 2013. The welfare state as an investment strategy: Denmark's flexicurity policies, in *The Oxford Handbook of Offshoring and Global Employment*, edited by A. Bardhan, D.M. Jaffee and C.A. Kroll. New York: Oxford University Press, 159–79.

Daily, B.F., Bishop, J.W. and Massoud, J.A. 2012. The role of training and empowerment in environmental performance: A study of the Mexican maquiladora industry. *International Journal of Operations and Production Management* 32(5), 631–47.

Daniels, S. 2010. Geographical imagination. *Transactions* 36(2), 182–7.

Das, G. and Han, H. 2013. Trade in middle products between South Korea and China, in *The Oxford Handbook of Offshoring and Global Employment*, edited by A. Bardhan, D.M. Jaffee and C.A. Kroll. New York: Oxford University Press, 276–310.

Davoudi, S. 2012. The legacy of positivism and the emergence of interpretative tradition in spatial planning. *Regional Studies* 46(4), 429–41.

Dawkins, R. 1989. *The Selfish Gene*. Oxford: Oxford University Press.

De Meyer, A. 1992. Management of international R&D operations, in *Technology Management and International Business*, edited by O. Granstrand, L. Håkanson and S. Sjölander. Chichester: Wiley, 162–79.

De Propris, L. and Hamdouch, A. 2013. Editorial: Regions as knowledge and innovative hubs. *Regional Studies* 47(7), 997–1000.

Delphi 2013. *Testing Services*. [Online]. Available at: http://delphi.com/manu
facturers/testing-services/mexico-technical-center/mtc-fat-lab/ [accessed: 23
February 2013].

Denzau, A.T. and North, D.C. 1994. Shared mental models: Ideologies and
institutions. *Kyklos* 47(1), 3–31.

Deschryvere, M. and Ali-Yrkkö, J. 2013. The impact of overseas R&D on
domestic R&D employment, in *The Oxford Handbook of Offshoring and
Global Employment*, edited by A. Bardhan, D.M. Jaffee and C.A. Kroll. New
York: Oxford University Press, 180–206.

Djelic, M.L. and Quack, S. 2008. Institutions and transnationalisation, in
Handbook of Organisational Institutionalism, edited by R. Greenwood,
C. Oliver, R. Suddaby and K. Sahlin-Andersson. Los Angeles: Sage, 299–323.

Domáński, B., Guzik, R., Gwosdz, K. and Dej, M. 2013. The crisis and beyond:
The dynamics and restructuring of automotive industry in Poland. *International
Journal of Technology and Management* 13(2), 151–66.

Dosi, G. 2012. A note on information, knowledge and economic theory, in
Handbook of Knowledge and Economics, edited by R. Arena, A. Festré and
L. Lazaric. Cheltenham and Northampton: Edward Elgar, 167–82.

Dühr, S. and Müller, A. 2012. The role of spatial data and spatial information in
strategic spatial planning. *Regional Studies* 46(4), 423–8.

Dunning, J.H. 2000. The eclectic paradigm as an envelope for economic and
business theories of MNE activity. *International Business Review* 9(2), 163–90.

Dutraive, V. 2012. The pragmatist view of knowledge and beliefs in institutional
economics: The significance of habits of thought, transactions and institutions
in the conception of economic behaviour, in *Handbook of Knowledge and
Economics*, edited by R. Arena, A. Festré and L. Lazaric. Cheltenham and
Northampton: Edward Elgar, 99–120.

Ebersberger, B. and Herstad, S.J. 2012. Go abroad or have strangers visit?
On organizational search spaces and local linkages. *Journal of Economic
Geography* 12(1), 273–97.

Edler, J., Meyer-Krahmer, F. and Reger, G. 2002. Changes in the strategic
management of technology: results of a global benchmarking study. *R&D
management* 32(2), 149–64.

Egidi, M. 2012. The cognitive explanation of economic behavior: From Simon
to Kahnemann, in *Handbook of Knowledge and Economics*, edited by
R. Arena, A. Festré and L. Lazaric. Cheltenham and Northampton: Edward
Elgar, 183–210.

Eisenhardt, K.M. and Martin, J.A. 2000. Dynamic capabilities: What are they?
Strategic Management Journal 21(10/11), 1105–21.

Ellegård, K. 1983. *Människa – Produktion: Tijdsbilder av ett Produktionssystem*.
Gothenburg: Göteborgs Universitets Geografiska Institutioner.

Ellegård, K. 1989. *Akrobatik i Tidens Väv: En Dokumentation av Projekteringen
av Volvos Bilfabrik i Uddevalla*. Gothenburg: Choros.

Entre Líneas 2013. *Inventa Conalep de Juárez Sistema Antirrobo de Autos. [Conalep de Juárez Invented an Anti-theft system].* [Online]. Available at: http://www. entrelineas.com.mx/notas.php?id_n=191575 [accessed: 13 September 2013].

Ernste, H. 2003. About unlearning and learning regions, in *Economic Geography of Higher Education. Knowledge Infrastructure and Learning Regions*, edited by R. Rutten, F. Boekema and E. Kuipers. London and New York: Routledge, 110–26.

ETUI (European Trade Union Institute) 2013a. *Facts and Figures*. [Online]. Available at: http://www.worker-participation.eu/European-Works-Councils/ Facts-Figures [accessed: 25 July 2013].

ETUI (European Trade Union Institute) 2013b. *How to Make your Meetings more Successful.* Brussels: ETUI. [Online]. Available at: http://www.worker-participation.eu/About-WP/Publications/Manual-for-European-Workers-Representatives-01-How-to-make-your-meetings-more-successful [accessed: 25 July 2013].

Faulconbridge, J. and McNeill, J. 2010. Geographies of space design. *Environment and Planning A* 42(12), 2820–23.

Fenton, C. and Langley, A. 2012. Strategy as practice and the narrative turn. *Organizational Studies* 32(9), 1171–96.

Fernandez-Stark, K., Bamber, P. and Gereffi, G. 2013. Regional competitiveness in the Latin America offshore services value chain, in *The Oxford Handbook of Offshoring and Global Employment*, edited by A. Bardhan, D.M. Jaffee and C.A. Kroll. New York: Oxford University Press, 534–63.

Ferreira, P., Vieira, E. and Neira, I. 2013. Culture impact on innovation. Econometric analysis of European Countries, in *Innovation in Socio-Cultural Context*, edited by F. Adam and H. Westlund. New York and London: Routledge, 57–79.

Fischer, P. 2012. *Phänomenologische Soziologie. [Phenomenological Sociology].* Bielefeld: Transkript.

Fleming, P. and Jones, M.T. 2013. *The End of Corporate Social Responsibility: Crisis and Critique.* Los Angeles and London: Sage.

Fleury, A. and Humphrey, J. 1993. *Human Resources and the Diffusion and Adaptation of New Quality Methods in Brazilian Manufacturing*. Sussex: IDS.

Flick, U. 2009. *An Introduction into Qualitative Research*. London: Sage.

Florida, R. 2007. *The Flight of the Creative Class: The New Global Competition for Talent*. New York: HarperCollins.

Florida, R., Mellander, C. and Stolarick, K. 2008. Inside the black box of regional development: Human capital, the creative class and tolerance. *Journal of Economic Geography* 8(5), 615–48.

Florida, R., Mellander, C. and Stolarick, K. 2012. Geographies of scope: An empirical analysis of entertainment, 1970–2000. *Journal of Economic Geography* 12(1), 183–204.

Foray, D. 2012. The fragility of experiential knowledge, in *Handbook of Knowledge and Economics*, edited by R. Arena, A. Festré and L. Lazaric. Cheltenham and Northampton: Edward Elgar, 267–84.

Franz, M. 2010. The potential of collective power in a global production network: Unicome and Metro Cash and Carry in India. *Erdkunde* 64(3), 281–90.

Freeman, C. 1995. The national system of innovation in historical perspective. *Cambridge Journal of Economics* 19(1), 5–24.

Freiberg, K., Freiberg, J. and Dunston, D. 2011. *Hanovation: How a Little Car Can Teach the World to Think Big and Act Bold.* Nashville: Thomas Nelson.

Freyssenet, M. and Lung, Y. 2004. Car firms' strategies and practices in Europe, in *European Industrial Restructering in a Global Economy: Fragmentation and Relocation of Value Chains*, edited by M. Faust, U. Voskamp and V. Wittke. Göttingen: SOFI, 85–103.

Fröbel, F., Heinrichs, J. and Kreye, O. 1980. *The New International Division of Labour: Structural Unemployment in Industrialised Countries and Industrialisation in Developing Countries.* Cambridge: Cambridge University Press.

Fromhold-Eisebith, M. and Fuchs, M. 2012. Changing global-local dynamics of economic development? Coining the new conceptual framework of 'industrial transition', in *Industrial Transition: New Global-Local Patterns of Production, Work, and Innovation*, edited by M. Fromhold-Eisebith and M. Fuchs. Farnham and Burlington: Ashgate, 1–16.

Fuchs, M. 2001. *Transnationale Lernprozesse in Ciudad Juárez, Mexiko. [Transnational Processes of Learning in Ciudad Juárez, Mexico].* Duisburg: INEF.

Fuchs, M. 2003a. 'Learning' in automobile components supply companies: The maquiladora of Ciudad Juárez, Mexico, in *Knowledge, Learning, and Regional Development*, edited by V. Lo and E.W. Schamp. Münster: Lit, 107–30.

Fuchs, M. 2003b. Borders and the internationalisation of knowledge: Two examples from the automobile components supply sector in Poland, in *Borders and Economic Behaviour in Europe*, edited by G. van Vilsteren and E. Wever. Assen: van Gorcum, 43–61.

Fuchs, M. 2005. Internal networking in the globalising firm: The case of R&D allocation in German automobile component supply companies, in *Linking Industries Across the World. Processes of Global Networking*, edited by C. Alvstam and E.W. Schamp. Burlington and Aldershot: Ashgate, 127–46.

Fuchs, M. and Endres, A. 2007. Wissenserwerb durch Nähe? Das Beispiel der Volkswagenzulieferer in Mexiko. [Acquisition of knowledge by proximity? The example of the supply industries of Volkwagen in Mexico]. *Geographische Rundschau* 59(1), 14–19.

Fuchs, M. 2008a. Subsidiaries of multinational companies: Foreign locations gaining competencies? *Geography Compass* 3(2), 1962–73.

Fuchs, 2008b. Product upgrading and survival of foreign plants: The case of VW Navarra, in *International Business Geography: Case Studies of Corporate Firms*, edited by P. Pellenbarg and E. Wever. London and New York: Routledge, 216–33.

Fuchs, M. and Winter, J. 2008. Competencies in subsidiaries of multinational companies: The case of the automotive supply industry in Poland. *Zeitschrift für Wirtschaftsgeographie* 52(4), 209–20.

Fuchs, M. and Scharmanski, A. 2009. Counteracting path dependencies: 'Rational' investment decisions in the globalizing commercial property market. *Environment and Planning A* 41(11), 2724–40.

Fuchs, M. and Meyer, D. 2010. Trade unions as learning organizations? The challenge of attracting temporary staff, in *Missing Links in Labour Geographies*, edited by A.C. Bergene, S.B. Endresen and H. Knutsen. Farnham and Burlington: Ashgate, 99–111.

Fuchs, M. 2012. Deutungsmuster als Beitrag zur wirtschaftsgeographischen Diskussion über Wissen und Institutionen. [Patterns of interpretation as a contribution to the debate about knowledge and institutions in economic geography]. *Geographische Zeitschrift* 100(2), 65–83.

Fuchs, M. and Kempermann, H. 2012. Flexible specialization: Thirty years after the 'second industrial divide': Lessons from the German mechanical engineering industry in the crisis 2008 to 2010, in *Industrial Transition: New Global-Local Patterns of Production, Work, and Innovation*, edited by M. Fromhold-Eisebith and M. Fuchs. Burlington and Farnham: Ashgate, 65–79.

Fuchs, M. and Fromhold-Eisebith, M. 2012. Conclusion: Towards a refined conceptualization of the industrial transition approach in a global-local context, in *Industrial Transition: New Global-Local Patterns of Production, Work, and Innovation*, edited by M. Fromhold-Eisebith and M. Fuchs. Farnham and Burlington: Ashgate, 233–41.

Fulton, L. 2007. *The Forgotten Resource: Corporate Governance and Employee Board-Level Representation: The Situation in France, the Netherlands, Sweden and the UK*. Düsseldorf: Hans Böckler Foundation.

Gassmann, O. and von Zedtwitz, M. 1999. New concepts and trends in international R&D organization. *Research Policy* 28(2/3), 231–50.

Geppert, M. and Dörrenbächer, C. 2011. Politics and power in the multinational corporation: An introduction, in *Politics and Power in the Multinational Corporation: The Role of Institutions, Interests, and Identities*, edited by C. Dörrenbächer and M. Geppert. Cambridge and New York: Cambridge University Press, 3–38.

Gereffi, G. 1999. International trade and industrial upgrading in the apparel commodity chain. *Journal of International Economics* 48(1), 37–70.

Gereffi, G., Humphrey, J. and Sturgeon, T. 2005. The governance of global value chains. *Review of International Political Economy* 12(1), 78–104.

Gertler, M. 2008. Buzz without being there? Communities of practice in context, in *Community: Economic Creativity and Organisation*, edited by A. Amin and J. Roberts. Oxford: Oxford University Press, 203–26.

Gibbons, M. 2003. A new mode of knowledge production, in *Economic Geography of Higher Education: Knowledge Infrastructure and Learning Regions*, edited by R. Rutten, F. Boekema and E. Kuipers. London and New York: Routledge, 229–42.

Giddens, A. 1984. *The Constitution of Society: Outline of the Theory of Structuration*. Cambridge and Oxford: Polity Press.

Giddens, A. 1993. *New Rules of Sociological Method: A Positive Critique of Interpretative Sociologies*. Second Edition. Cambridge and Oxford: Polity Press and Blackwell.

Girndt, C. 2012. Tochterfirmen zu einem Aktivposten machen. [To turn subsidiaries in assets]. *Mitbestimmung* 58(3), 22–5.

Glassner, V. 2012. Transnational collective bargaining in national systems of industrial relations, in *Transnational Collective Bargaining at Company Level: A New Component of European Industrial Relations?*, edited by I. Schömann, R. Jagodzinski, G. Boni, S. Clauwaert, V. Glassner and T. Jaspers. Brussels: ETUI, 77–115.

Glückler, J. 2011. Islands of expertise. Global knowledge transfer in a technology service firm, in *Beyond Territory: Dynamic Geographies of Knowledge Creation, Diffusion, and Innovation*, edited by H. Bathelt, M.P. Feldman and D.F. Kogler. London and New York: Routledge, 207–26.

Gomory, R.E. and Baumol, W.J. 2013. On technical progress and the gains and losses from outsourcing, in *The Oxford Handbook of Offshoring and Global Employment*, edited by A. Bardhan, D.M. Jaffee and C.A. Kroll. New York: Oxford University Press, 24–40.

Goodman, M.K., Boykoff, M.T. and Evered, K.T. 2008. Contentious geographies: Environmental knowledge, meaning, scale, in *Contentious Geographies: Environmental Knowledge, Meaning, Scale*, edited by M.K. Goodman, M.T. Boykoff and K.T. Evered. Aldershot and Burlington: Ashgate, 1–23.

Gordon, E.E. 2009. *Winning the Global Talent Showdown: How Businesses and Communities Can Partner to Rebuild the Jobs Pipeline*. San Francisco: Berrett-Koehler Publishers.

Grabher, G. 2002. Cool projects, boring institutions: Temporary collaboration in social context. *Regional Studies* 36(3), 205–14.

Gregory, D. 2007. Imaginative geographies, in *The Dictionary of Human Geography*, edited by D. Gregory, R. Johnson, G. Pratt, M. Watts and S. Whatmore. Oxford: Blackwell, 369–71.

Greifenstein, R. and Kißler, L. 2012. *Co-determination in the Focus of Social Research – 1952–2010*. Düsseldorf: Hans Böckler Foundation.

Guimón, J. 2008. *Government Strategies to Attract R&D-intensive FDI*. [Online]. Available at: http://www.oecd.org/investment/globalforum/403108 56.pdf [accessed: 16 July 2013].

Gutierrez, J.J. 2012. Innovation in low- and medium-technology manufacturing: The role of networks and non-R&D inputs, in *The Regional Economics of Knowledge and Talent: Local Advantage in a Global Context*, edited by C. Karlsson, B. Johansson and R.R. Stough. Cheltenham and Northampton: Edward Elgar, 134–55.

Haakonsson, S.J., Ørberg Jensen, P.D. and Mudambi, S.M. 2013. A co-evolutionary perspective on the drivers of international sourcing of pharmaceutical R&D to India. *Journal of Economic Geography* 13(4), 677–700.

Hacking, I. 1999. *The Social Construction of What?* Cambridge and London: Harvard University Press.

Håkanson, L. 1995. Learning through acquisitions: Management and integration of foreign R&D laboratories. *International Studies of Management and Organization* 25(1–2), 121–57.

Harrison, J. 2013. Configuring the new 'regional world': On being caught between territory and networks. *Regional Studies* 47(1), 55–74.

Harvey, D. 1990. *The Condition of Postmodernity: An Enquiry into the Origins of Cultural Change.* Cambridge: Blackwell.

Hassink, R. 2005. How to unlock regional economies from path dependencies? *European Planning Studies* 13(4), 521–35.

Hassink, R. and Klaerding C. 2012. The end of the learning region as we knew it: Towards learning in space. *Regional Studies* 46(8), 1055–66.

Hauser-Ditz, A., Hertwig, M., Pries, L. and Rampeltshammer, L. 2010. *Transnationale Mitbestimmung? Zur Praxis Europäischer Betriebsräte in der Automobilindustrie. [Transnational Co-determination? The Practice of European Works Councils in the Automobile Industry].* Frankfurt: Campus.

Hayek, F.A. 1937. Economics and knowledge. *Economica* 4(13), 96–105.

Hayek, F.A. 1945. The use of knowledge in society. *American Economic Review* 25(4), 519–30.

Hayter, R. and Le Heron, R. 2002. Industrialization, techno-economic paradigms and the environment, in *Knowledge, Industry and Environment: Institutions and Innovation in a Territorial Perspective*, edited by R. Hayter and R. Le Heron. Aldershot and Burlington: Ashgate, 11–30.

Healy, A. and Morgan, K. 2012. Spaces of innovation: Learning, proximity and the ecological turn. *Regional Studies* 46(8), 1041–53.

Hecker, A. 2012. Knowledge beyond the individual? Making sense of a notion of collective knowledge in organization theory. *Organization Studies* 33(3), 423–45.

Helbrecht, I. 2011. Die Welt als Horizont: Zur Produktion globaler Expertise in der Weltgesellschaft [The world as horizon: The production of global expertise in the global society], in *Räume der Wissensarbeit. Zur Funktion von Nähe und Distanz in der Wissensökonomie [Spaces of Knowledge Work. The Role of Proximity and Distance in the Knowledge Economy]*, edited by O. Ibert and H.J. Kujath. Wiesbaden: VS Publisher, 103–24.

Henderson, J., Dicken, P., Hess, M., Coe, N. and Yeung, H. 2002. Global production networks and the analysis of economic development, *Review of International Political Economy* 9(3), 436–64.

Hinchliffe, S. 2000. Performance and experimental knowledge: Outdoor management training and the end of epistemology. *Environment and Planning D* 18(5), 575–95.

Hirsch-Kreinsen, H. 2008. 'Low-tech' innovations, *Industry and Innovation* 15(1), 19–43.

Hirsch-Kreinsen, H. 2009. Innovative Arbeitspolitik im Maschinenbau? [Innovative labour market policy in engine building industry?]. Dortmund: University Press.

Hirsch-Kreinsen, H. 2012. Einfache Produkte intelligent produzieren [Intelligent production of simple products], in *Sozialwissenschaftliche Beiträge zur Produktionsforschung, [Contibutions in Social Sciences to Research about Production]*, edited by H. Hirsch-Kreinsen, G. Lay and J. Abel. Stuttgart: Fraunhofer Publisher, 129–37.

Hirsch-Kreinsen, H., Lay, G. and Abel, J. 2012. Die Entwicklung sozialwissenschaftlicher Beiträge zur Produktionsforschung [The development of contributions in social sciences to research about production], in *Sozialwissenschaftliche Beiträge zur Produktionsforschung, [Contibutions in Social Sciences to Research about Production]*, edited by H. Hirsch-Kreinsen, G. Lay and J. Abel. Stuttgart: Fraunhofer Publisher, 9–24.

Holm, U., Johanson, J. and Thilenius, P. 1995. Headquarters' knowledge of subsidiary network contexts in the multinational corporation. *International Studies of Management and Organization* 25(1–2), 97–117.

Homm, S. and Bohle, H.-G. 2012. India's Shenzhen – A miracle? Critical reflections on the new economic geography, with empirical evidence from peri-urban Chennai. *Erdkunde* 66(4), 281–94.

Honeywell 2013. *Mexico*. [Online]. Available at: http://honeywell.com/World wide/Pages/mexico-en.aspx [accessed: 16 August 2013].

Horgos, D. 2013. The sector bias of offshoring: Empirical importance for labor-market implications, in *The Oxford Handbook of Offshoring and Global Employment*, edited by A. Bardhan, D.M. Jaffee and C.A. Kroll. New York: Oxford University Press, 100–121.

Hosseini, S.A.H. 2010. *Alternative Globalizations: An Integrative Approach to Studying Dissident Knowledge in the Global Justice Movement*. London and New York: Routledge.

Howells, J. 2012. The geography of knowledge: Never so close but never so far apart. *Journal of Economic Geography* 5(12), 1003–20.

Huber, F. 2012. On the role and interrelationship of spatial, social and cognitive proximity: Personal knowledge relationships of R&D workers in the Cambridge information technology cluster. *Regional Studies* 46(9), 1169–82.

Hudson, R. 1999. The learning economy, the learning firm and the learning region: A sympathetic critique of the limits to learning. *European Urban and Regional Studies* 6(1), 59–72.

Huf Hülsbeck and Fürst 2012. *Huf Hülsbeck and Fürst: Seit 104 Jahren im Dienste der Automobilindustrie. [Huf Hülsbeck and Fürst: Since 104 Years Working for Automobile Industry].* [Online]. Available at: http://www.huf-group.com/home/openCms.nsf/DEUTSCH%20-%202012%20Oktober%20Press%20Info%20Huf%20Portrait.pdf [accessed: 4 April 2013].

Huf Hülsbeck and Fürst 2013. Welcome to Huf. [Online]. Available at: http://www.huf-group.com/index.php?id=90 [accessed: 4 April 2013].

Ibert, O. 2007. Towards a geography of knowledge creation: The ambivalences between 'knowledge as an object' and 'knowing in practice'. *Regional Studies* 41(1), 103–14.

Ibert, O. 2010. Relational distance: Sociocultural and time-spatial tensions in innovation practices. *Environment and Planning A* 42(1), 187–204.

Ibert, O. and Kujath, H.J. 2011. Wissensarbeit aus räumlicher Perspektive: Begriffliche Grundlagen und Neuausrichtungen im Diskurs [Knowledge work from a spatial view: Terminological foundation and reframing in the discourse], in *Räume der Wissensarbeit. Zur Funktion von Nähe und Distanz in der Wissensökonomie, [Spaces of Knowledge Work: The Role of Proximity and Distance in the Knowledge Economy]*, edited by O. Ibert and H.J. Kujath. Wiesbaden: VS Publisher, 9–46.

ILO 2013. *The Benefits of International Labour Standards*. [Online]. Available at: http://www.ilo.org/global/standards/introduction-to-international-labour-standards/the-benefits-of-international-labour-standards/lang--en/index.htm [accessed: 27 July 2013].

INADET 2013a. *CENALTEC*. [Online]. http://inadet.com.mx/cenaltec/ [accessed: 13 September 2013].

INADET 2013b. *CENALTEC*. [Online]. http://inadet.com.mx/clientes-distingui dos-cenaltec-juarez/ [accessed: 13 September 2013].

INEGI 2013. *Estadística [Statistics]*. [Online]. Available at: http://www.inegi.org. mx/ [accessed: 24 May 2013].

Jagodzinski, R. 2012. Transnational collective bargaining: A literature review, in *Transnational Collective Bargaining at Company Level: A New Component of European Industrial Relations?*, edited by I. Schömann, R. Jagodzinski, G. Boni, S. Clauwaert, V. Glassner and T. Jaspers. Brussels: ETUI, 19–76.

Jessop, B. and Sum, N.-L. 2006. *Beyond the Regulation Approach: Putting Capitalist Economies in their Place*. Cheltenham: Edward Elgar.

Jessop, B., Moulaert, F., Hulgård, L. and Hamdouch, A. 2013. Social innovation research: A new stage in innovation analysis? in *The International Handbook on Social Innovation: Collective Action, Social Learning and Transdisciplinary Research*, edited by F. Moulaert, D. MacCallum, A. Mehmood and A. Hamdouch. Cheltenham and Northampton: Edward Elgar, 110–30.

Johanson, J. and Vahlne, J.E. 1977. The internationalization process of the firm: A model of knowledge development and increasing foreign market commitments. *Journal of International Business Studies* 8(1), 23–32.

Johnson-Laird, P. 1983. *Mental models*. Cambridge: Harvard University Press.

Jones, A. 2008. Beyond embeddedness: Economic practices and the invisible dimensions of transnational business activity, *Progress in Human Geography* 32(1), 71–88.

Jones, M. and Woods, M. 2012. New localities. *Regional Studies* 47(1), 29–42.

Jöns, H., Livingstone, D.N. and Meusberger, P. 2010. Interdisciplinary geographies of science, in *Geographies of Science*, edited by P. Meusburger, D.N. Livingstone and H. Jöns. Heidelberg, London and New York: Springer, ix–xvii.

Jullien, B. and Pardi, T. 2013. Structuring new automotive industries, restructuring old automotive industries and the new geopolitics of the global automotive sector. *International Journal of Technology and Management* 13(2), 96–113.

Jürgens, U. and Krzywdzinski, M. 2013. Breaking off from national bounds: Human resource management practices of national players in the BRIC countries. *International Journal of Technology and Management* 13(2), 114–33.

Justesen, L. and Mouritsen, J. 2009. The triple visual. Translations between photographs, 3-D visualizations and calculations. *Accounting, Auditing and Accountability Journal* 22(6), 973–90.

Kahn, K.B. and McDonough, E.F. 1997. Marketing's integration with R&D and manufacturing: A cross-regional analysis. *Journal of International Marketing* 5(1), 51–76.

Kant, I. 1787/2011. *Kritik der reinen Vernunft. [Critique of Pure Reason].* Köln: Anaconda.

Kaplinsky, R., Joffe, A. Kaplan, D. and Lewis, D. 1995. *Improving Manufacturing Performance in South Africa: Report of the Industrial Strategy Project.* Cape Town: UCT Press.

Karlsson, C. and Johansson, B. 2012. Knowledge, creativity and regional development, in *The Regional Economics of Knowledge and Talent: Local Advantage in a Global Context*, edited by C. Karlsson, B. Johansson and R.R. Stough. Cheltenham and Northampton: Edward Elgar, 27–62.

Karlsson, C., Johansson, B. and Stough, R.R. 2012. Introduction. Human capital and agglomeration, in *The Regional Economics of Knowledge and Talent: Local Advantage in a Global Context*, edited by C. Karlsson, B. Johansson and R.R. Stough. Cheltenham and Northampton: Edward Elgar, 1–24.

Kassner, K. 2003. Soziale Deutungsmuster: Über aktuelle Ansätze zur Erforschung kollektiver Sinnzusammenhänge [Social patterns of interpretation: About recent approaches to analyse collective contexts of meaning], in *Sinnformeln [Formula of Meaning]*, edited by S. Geideck and W.-A. Liebert. Berlin: De Gruyter, 37–57.

Kauffeld-Monz, M. and Fritsch, M. 2010. Who Are the Knowledge Brokers in Regional Systems of Innovation? A Multi-Actor Network Analysis. *Regional Studies* 47(5), 669–85.

Keller, R. 2011. *Wissenssoziologische Diskursanalyse. [Discourse Analysis in Sociology of Knowledge].* Wiesbaden: VS Publisher.

Kern, H. and Schumann, M. 1987. Limits of the division of labour: New production and employment concepts in West German industry. *Economic and Industrial Democracy* 8(2), 151–70.

Kiella, M.L. and Golhar, D.Y. 1997. Total quality management in an R&D environment. *International Journal of Operations and Production Management* 17(2), 184–98.

Kijkuit, B. and van den Ende, J. 2010. With a little help from our colleagues: A longitudinal study of social networks for innovation. *Organization Studies* 31(4), 451–79.

Kinkel, S. 2012. Trends in production relocation and backshoring activities: Changing patterns in the course of the global economic crisis. *International Journal of Operations and Production Management* 32(6), 696–720.

Kirchhoff 2012. On course for further growth. *K Mobil* 17(80), 48–9.

Kirchhoff 2013a. *Kirchhoff.* [Online]. Available at: http://www.kirchhoff-gruppe. de/ [accessed: 23 July 2013].

Kirchhoff 2013b. *200 Jahre Witte. [200 Years Witte].* [Online]. Available at: http:// www.kirchhoff-gruppe.de/cms/unternehmensgruppe_historie [accessed: 23 July 2013].

Kirner, E., Som, O. and Jäger, A. 2009. *Vernetzungsmuster und Innovationsverhalten von nicht forschungsintensiven Betrieben. [Patterns of Networking and Innovation Behaviour of Companies Without Intensive Research].* Stuttgart: Fraunhofer Publisher.

Knoben, J. and Oerlemans, L.A.G. 2012. Configurations of inter-organizational knowledge links: Does spatial embeddedness still matter? *Regional Studies* 46(8), 1005–21.

Knorr-Cetina, K. 1999. *Epistemic Cultures: How the Sciences make Knowledge.* Cambridge and London: Harvard University Press.

Kopinak, K. 1996. *Desert Capitalism: Maquiladoras in North America's Western Industrial Corridor.* Tucson: University of Arizona Press.

Kriegesmann, B., Kley, T. and Kublik, S. 2010. Innovationstreiber betriebliche Mitbestimmung? [Co-determination as driving force for innovation?] *WSI Mitteilungen* 63(2), 71–8.

Krikorian, G. and Kapczynski, A. 2010. Preface, in *Access to Knowledge in the Age of Intellectual Property*, edited by G. Krikorian and A. Kapczynski. New York: Zone Books, 9–14.

Kroll, H. and Tagscherer, U. 2009. *Chinese Regional Innovation Systems in Times of Crisis: The Case of Guangdong.* Karlsruhe: Fraunhofer ISI.

Kuhn, T.S. 1962/2012. *The Structure of Scientific Revolutions.* Chicago and London: University of Chicago Press.

Kuula, M., Putkiranta, A. and Toivanen, J. 2012. Coping with the change: A longitudinal study into the changing manufacturing practices. *International Journal of Operations and Production Management* 32(2), 106–20.

Lagendijk, A. 2001. Scaling knowledge production: How significant is the region? in *Knowledge, Complexity and Innovation Systems*, edited by M.M. Fischer and J. Fröhling. Berlin: Springer, 79–100.

Lau, I.Y.-M., Chiu, C.-Y. and Lee, S.-L. 2001. Communication and shared reality: Implications for the psychological foundations of culture. *Social Cognition* 19(3), 350–71.

Lee, C.-K. and Saxenian, A. 2008. Coevolution and coordination: A systemic analysis of the Taiwanese information technology industy. *Journal of Economic Geography* 8(2), 157–80.

Lema, R., Quadros, R. and Schmitz, H. 2012. *Shifts in Innovation Power to Brazil and India: Insights from the Auto and Software Industries*. Brighton: Institute of Development Studies.

Leydesdorff, L. 2006. *The Knowledge-based Economy: Modeled, Measured, Simulated*. Boca Raton: Universal Publishers.

Li, L. 2005. The effects of trust and shared vision on inward knowledge transfer in subsidiaries' intra- and inter-organizational relationships. *International Business Review* 14(1), 77–95.

Lipietz, A. 1986. *Mirages et Miracles: Problèmes de l'Industrialisation dans le Tiers Monde. [Mirages and Miracles: Problems of Industrialisation in the Third World]*. Paris: La Découverte.

Lipietz, A. 1987. *Mirages and Miracles. The Crisis of Global Fordism*. London: Verso.

Lippert, I. and Jürgens, U. 2012. *Corporate Governance und Arbeitnehmerbeteiligung in den Spielarten des Kapitalismus. [Corporate Governance and Employees' Representation in the Varieties of Capitalism]*. Berlin: Edition Sigma.

Livingstone, D.N. 2003. *Putting Science in its Place: Geographies of Scientific Knowledge*. Chicago and London: University of Chicago Press.

Lockwood, N.R. 2010. *Corporate India and HR Management: Creating Talent Pipelines, Leadership Competencies, and Human Resources*. Alexandria: Society For Human Resource Management.

Loewenthal, D. 1961. Geography, experience, and imagination: Towards a geographical epistemology. *Annals of American Geographers* 51(3), 241–60.

Lorenz, E. and Lundvall, A. 2006. Understanding European systems of competence building, in *How Europe's Economies Learn: Coordinating Competing Models*, edited by E. Lorenz, B.-Å. Lundvall. Oxford: Oxford University Press, 1–25.

Lundvall, B.-Å. 1992. *National Innovation Systems: Towards a Theory of Innovation and Inter-active Learning*. London: Pinter.

MacKinnon, D., Cumbers, A., Pike, A., Birch, K. and McMaster, R. 2009. Evolution in economic geography: Institutions, political economy, and adaptation. *Economic Geography* 85(2), 129–50.

Malecki, E.J. 2012. Regional social capital: Why it matters. *Regional Studies* 46(8), 1023–39.

Malish, C.M. and Vigneswara Ilavarasan, P. 2011. Social exclusion in information capitalism: A study of online recruitment advertisements in the Indian software industry, in *Global Knowledge Work: Diversity and Relational Perspectives*, edited by K. Nicolopoulou, M. Karataş-Özkan, A. Tatli and J. Taylor. Cheltenham and Northampton: Edward Elgar, 114–39.

Malmberg, A. and Maskell, P. 1999. The competitiveness of firms and regions: 'Ubiquitification' and the importance of localized learning. *European Urban and Regional Studies* 6(1), 9–25.

Mannheim, K. 1925. Das Problem einer Soziologie des Wissens [The problem of a sociology of knowledge]. *Archiv für Sozialwissenschaft und Sozialpolitik* 53, 577–652.

Manning, S., Sydow, J. and Windeler, A. 2012. Securing access to lower-cost talent globally: The dynamics of active embedding and field structuration. *Regional Studies* 46(9), 1201–18.

Markusen, A. 1996. Sticky places in a slippery space: A typology of industrial districts. *Economic Geography* 72(3), 293–313.

Marrocu, E. and Paci, R. 2012. Education and creativity: What matters most for economic performance? *Economic Geography* 88(4), 369–401.

Marston, S.A. and de Leeuw, S. 2013. Creativity and geography: Toward a politicized intervention. *Geographical Review* 103(2), iii–xxvi.

Martin, R. and Sunley, P. 2006. Path dependence and regional economic evolution. *Journal of Economic Geography* 6(4), 395–437.

Martin, R. and Moodysson, J. 2011. Innovation in symbolic industries: The geography and organization of knowledge sourcing. *European Planning Studies* 19(7), 1183–203.

Martin, R. 2012a. *Knowledge Bases and the Geography of Innovation.* Lund: Lund University Press.

Martin, R. 2012b. Measuring knowledge basis in Swedish regions. *European Planning Studies* 20(9), 1569–82.

Martinez-Fernandez, C. and Miles, I. 2011. Knowledge-intensive service activities: Integrating knowledge for innovation, in *The Knowledge Economy at Work: Skills and Innovation in Knowledge-intensive Service Activities*, edited by C. Martinez-Fernandez, I. Miles and T. Weyman. Cheltenham and Northampton: Edward Elgar, 1–19.

Massey, D. 1984. *Spatial Divisions of Labour: Social Structures and the Geography of Production.* London: Methuen.

McKinsey 2011a. *Beyond Expats: Better Managers for Emerging Markets.* [Online]. Available at: http://www.mckinsey.com/insights/organization/beyond _expats_better_managers_for_emerging_markets [accessed: 4 September 2013].

McKinsey 2011b. *Survey: R&D Strategies in Emerging Economies: McKinsey Global Survey Results.* [Online]. Available at: http://www.mckinsey.com/ insights/operations/r_and_38d_strategies_in_emerging_economies_mckin sey_global_survey_results [accessed: 4 September 2013].

McKinsey 2012a. *How Multinationals Can Attract the Talent They Need.* [Online]. Available at: http://www.mckinsey.com/insights/organization/how_ multinationals_can_attract_the_talent_they_need [accessed: 4 September 2013].

McKinsey 2012b. *How Multinationals Can Win in India.* [Online]. Available at: http://www.mckinsey.com/insights/winning_in_emerging_markets/how_ multinationals_can_win_in_india [accessed: 4 September 2013].

McKinsey 2012c. *Organizing for an Emerging World.* [Online]. Available at: http:// www.mckinsey.com/insights/organization/organizing_for_an_emerging_ world [accessed: 4 September 2013].

McKinsey 2013. *Urban world: The shifting global business landscape.* [Online]. Available at: http://www.mckinsey.com/Insights/Urbanization/Urban_world_ The_shifting_global_business_landscape?cid=other-eml-alt-mgi-mck-oth-1310 [accessed: 3 October 2013].

Messner, D. 2002. World society: Structures and trends, in *Global Trends and Global Governance*, edited by P. Kennedy, D. Messner and F. Nuscheler. London and Sterling: Pluto Press: 22–64.

Meyer, R.E., Höllerer, M.A., Jancsary, D. and van Leeuwen, T. 2013. The visual dimension in organizing, organization and organization research: Core ideas, current developments, and promising avenues. *The Academy of Management Annals* 7(1), 487–553.

MFI International 2012. *High-Tech Training for Companies Looking at Manufacturing in Mexico.* Available at: http://info.mfiintl.com/blog/bid/1272 82/High-Tech-Training-for-Companies-Looking-at-Manufacturing-in-Mexico [accessed: 13 September 2013].

Michie, J. and Sheehan, M. 1999. HRM practices, R&D expenditure and innovative investment: Evidence from the UK's 1990 workplace industrial relations survey (WIRS). *Industrial and Corporate Change* 8(2), 211–34.

Michie, J. 2011. Globalisation: Introduction and overview, in *The Handbook of Globalisation*, edited by J. Michie. Second Edition. Cheltenham and Northampton: Edward Elgar, 1–16.

Mills, C.W. 1956/2000. *The Power Elite.* New York and Oxford: Oxford University Press.

Mittelman, J. 2004. *Whither Globalization? The Vortex of Knowledge and Ideology.* London and New York: Routledge.

Mohammed, S. and Dumville, B.C. 2001. Team mental models in a team knowledge framework: Expanding theory and measurement across disciplinary boundaries. *Journal of Organizational Behavior* 22(2), 89–106.

Molitor, C. 2012. Alternativvorschläge – Nicht erwünscht! Die Investoren aus den BRIC-Staaten. [Alternative proposals unwanted. The investors of the BRIC countries]. *Mitbestimmung* 58(3), 29–33.

Molitor, C. 2013a. Abschied von Spritfressern. [Farewell to gaz guzzlers]. *Mitbestimmung* 59(3), 34.

Molitor, C. 2013b. Vom Sorgenkind zum Vorzeigestandort. [From problem child to flagship location]. *Mitbestimmung* 59(3), 35.

Moodysson, J. 2011. Principles and practices of knowledge creation: On the organization of 'buzz' and 'pipelines' in life science communities. *Economic Geography* 84(4), 449–69.

Morgan, G. 2011. Reflections on the macro-politics of micro-politics, in *Politics and Power in the Multinational Corporation: The Role of Institutions, Interests, and Identities*, edited by C. Dörrenbächer and M. Geppert. Cambridge and New York: Cambridge University Press, 415–36.

Moulaert, F. and Nussbaumer, J. 2005. The social region. Beyond the territorial dynamics of the learning economy. *European Urban and Regional Studies* 12(1), 45–64.

Myrdal, G. 1957. *Economic Theory and Underdeveloped Regions*. London: Gerald Duckworth.

Nadvi, K. 2008. Global standards, global governance and the organization of global value chains. *Journal of Economic Geography* 8(3), 323–43.

Nathan, D. and Sarkar, S. 2013. *Innovation and Upgrading in Global Production Networks*. Manchester: The University of Manchester, Working Papers Capturing the Gains 23. [Online]. Available at: www.capturingthegains.org/publications/workingpapers/wp_201323.htm [accessed: 22 July 2013].

Navas-Alemán, L. 2011. The impact of operating in multiple value chains for upgrading: The case of the Brazilian furniture and footwear industries. *World Development* 39(8), 1386–97.

Nelson, R. 1993. *National Innovation Systems: A Comparative Analysis*. New York and Oxford: Oxford University Press.

Nelson, R. 2009. Routines as technologies and as organizational capabilities, in *Organizational Routines: Advancing Empirical Research*, edited by M.C. Becker and N. Lazaric. Cheltenham and Northampton: Edward Elgar, 11–25.

NIST (National Institute of Standards and Technology) 2010: *Definition of Scope: Manufacturing-related R&D*. [Online]. Available at: http://www.nist.gov/tpo/sbir/mfgdefinition.cfm [accessed: 16 July 2013].

Nonaka, I. and Takeuchi, H. 1995. *The Knowledge Creating Company: How Japanese Companies Create the Dynamics of Innovation*. Oxford and New York: Oxford University Press.

Nooteboom, B. 2012. Embodied cognition, organization and innovation, in *Handbook of Knowledge and Economics*, edited by R. Arena, A. Festré and L. Lazaric. Cheltenham and Northampton: Edward Elgar, 339–68.

North, D.C. 1991. Institutions. *The Journal of Economic Perspectives* 5(1), 97–112.

Novelli, M. and Ferus-Comelo, A. 2010. Globalization and knowledge production in labour movements, in *Globalization, Knowledge and Labour: Education for Solidarity within Spaces of Resistance*, edited by M. Novelli and A. Ferus-Comelo. London and New York: Routledge, 49–63.

O'Connell, D., Hickerson, K., Pillutla, A. 2012. Organizational visioning: An integrative review. *Group and Organization Management* 37(5), 655–85.

OECD 1996. *The Knowledge-based economy.* Paris: OECD Publishing. [Online]. Available at: http://www.oecd.org/science/sci-tech/1913021.pdf [accessed: 22 May 2013].

OECD 2008. *Research and Development: Going Global.* [Online]. Available at: http://www.oecd.org/science/sci-tech/41090260.pdf [accessed: 5 September 2013].

OECD 2009. *Innovation in Firms. A Microeconomic Perspective.* Paris: OECD Publishing.

OECD 2010. *The Paso del Norte Region, Mexico, and the United States*. [Online]. Available at: http://www.oecd.org/mexico/45820961.pdf [accessed: 24 September 2013].

OECD 2011. *ISIC Rev. 3: Technology Intensity Definition, Classification of Manufacturing Industries into Categories Based on R&D Intensities*. [Online]. Available at: http://www.oecd.org/science/inno/48350231.pdf [accessed: 27 February 2013].

OECD 2012a. *OECD Science, Technology and Industry Outlook 2012*. [Online]. Available at: http://www.oecd.org/sti/sti-outlook-2012-highlights. pdf [accessed: 25 May 2013].

OECD 2012b. *OECD Indicators. Education at a Glance*. Paris: OECD Publishing.

OECD 2012c. *KEI and KI Indexes* [Online]. Available at: http://info.worldbank. org/etools/kam2/KAM_page5.asp [accessed: 21 July 2014].

OECD 2012d. *KEI and KI Indexes (KAM 2012)*. [Online]. Available at: http://info. worldbank.org/etools/kam2/KAM_page5.asp [accessed: 17 September 2013].

OECD 2013a. Expenditure on R&D, in *OECD Factbook 2013: Economic, Environmental and Social Statistics*. Paris: OECD Publishing. [Online]. Available at: http://www.oecd-ilibrary.org/sites/factbook-2013-en/08/01/01/ index.html?contentType=&itemId=/content/chapter/factbook-2013-60-en&containerItemId=/content/serial/18147364&accessItemIds=&mimeType= text/html [accessed: 25 May 2013].

OECD 2013b. Researchers, in *OECD Factbook 2013: Economic, Environmental and Social Statistics*. Paris: OECD Publishing. [Online]. Available at: http://www.oecd-ilibrary.org/sites/factbook-2013-en/08/01/02/index. html?contentType=&itemId=/content/chapter/factbook-2013-61-en&containerItemId=/content/serial/18147364&accessItemIds=&mimeType= text/html [accessed: 25 May 2013].

OECD 2013c. Patents, in *OECD Factbook 2013: Economic, Environmental and Social Statistics*. Paris: OECD Publishing. [Online]. Available at: http://www.oecd-ilibrary.org/sites/factbook-2013-en/08/01/03/index. html?contentType=&itemId=/content/chapter/factbook-2013-62-en&containerItemId=/content/serial/18147364&accessItemIds=&mimeType= text/html [accessed: 25 May 2013].

OECD 2013d. *Innovation: Economic Crisis and Weak Outlook Hit R&D*. [Online]. Available at: http://www.oecd.org/newsroom/innovationecon omiccrisisandweakoutlookhitrdsaysoecd.htm [accessed: 25. May 2013].

OECD 2013e. *Gross Domestic Expenditure on R&D* [Online]. Available at: http://www.oecd-ilibrary.org/sites/factbook-2013-en/08/01/01/index.html? contentType=&itemId=/content/chapter/factbook-2013-60-en&container ItemId=/content/serial/18147364&accessItemIds=&mimeType=text/html [accessed: 22 May 2013].

OECD 2013f. *Reseachers*. [Online]. Available at: http://www.oecd-ilibrary.org/ economics/oecd-factbook-2013/researchers_factbook-2013–61-en [accessed: 25 May 2013].

OECD 2013g. *Number of Patents.* [Online]. Available at: http://stats.oecd.org/ Index.aspx?DatasetCode=PATS_IPC [accessed: 25. May 2013].

Oevermann, U. 1973/2001a. Zur Analyse der Struktur von sozialen Deutungs-mustern. [About the analysis of the structure of social patterns of interpretation]. *Sozialer Sinn* 2(1), 3–33.

Oevermann, U. 2001b. Die Struktur von sozialen Deutungsmustern: Versuch einer Aktualisierung. [The structure of social patterns of interpretation: Approach for an update]. *Sozialer Sinn* 2(1), 35–81.

Oevermann, U. 2001c. *Strukturprobleme supervisorischer Praxis. [Structural Problems of the Practice of Supervision].* Frankfurt: Humanitas.

Oinas, P. 2000. Distance and learning: Does proximity matter? in *Knowledge, Innovation and Economic Growth: Theory and Practice of the Learning Region*, edited by F. Boekema, K. Morgan, S. Bakkers and R. Rutten. Cheltenham: Edward Elgar, 57–69.

Oshri, I., van Fenema, P.C. and Kotlarsky, J. 2008. Knowledge transfer in globally distributed team: The role of transactive memory, in *Knowledge Processes in Globally Distributed Contexts*, edited by I. Oshri, P.C. van Fenema and J. Kotlarsky. Basingstoke and New York: Palgrave Macmillan, 24–52.

OVTA 2013. *OVTA – Helping Businesses Develop Human Resources for Globalization.* [Online]. Available at: http://www.ovta.or.jp/en/ [accessed: 28 August 2013].

Parthasarathy, B. 2013. The changing character of Indian offshore ICT services provision, 1985–2010, in *The Oxford Handbook of Offshoring and Global Employment*, edited by A. Bardhan, D.M. Jaffee and C.A. Kroll. New York: Oxford University Press, 380–404.

Patalano, R. 2012. Imagination and perception as gateways to knowledge: The unexplored affinity between Boulding and Hayek, in *Handbook of Knowledge and Economics*, edited by R. Arena, A. Festré and L. Lazaric. Cheltenham and Northampton: Edward Elgar, 121–43.

Patel, P. and Pavitt, K. 1991. Large firms in the production of world's technology: An important case of non-globalization. *Journal of International Business Studies* 22(1), 1–21.

Pavlinek, P. 2012. The internationalization of corporate R&D and automotive industry R&D of East-Central Europe. *Economic Geography* 88(3), 297–310.

Pearce, C.L. 2004. The future of leadership: Combining vertical and shared leadership to transform knowledge work. *Academy of Management* 18(1), 47–57.

Peck, J. 1996. *Workplace: The Social Regulation of Labour Markets.* New York: Guilford Press.

Persson, M. and Åhlström, P. 2013. Product modularisation and organisational coordination. *International Journal of Technology and Management* 13(2), 55–74.

Phelps, N.A. and Fuller, C. 2000. Multinationals, intracorporate competition, and regional development. *Economic Geography* 76(3), 224–43.

Phelps, N.A. and Wood, A. 2006. Lost in translation? Local interests, global actors and inward investment regions. *Journal of Economic Geography* 6(4), 493–515.

Pietrobelli, C. and Saliola, F. 2008. Power relationships along the value chain: Multinational firms, global buyers and performance of local suppliers. *Cambridge Journal of Economics* 32(6), 947–62.

Pilz, M. (ed.) 2012. *The Future of Vocational Education and Training in a Changing World.* Wiesbaden: VS Publisher.

Piore, M. and Sabel, C.F. 1984. *The Second Industrial Divide.* New York: Basic Books.

Plank, L. and Staritz, L. 2013. *'Precarious Upgrading' in Electronics Global Production Networks in Central and Eastern Europe: The Cases of Hungary and Romania.* Manchester: The University of Manchester, Working Papers Capturing the Gains 31. [Online]. Available at: www.capturingthegains.org/pdf/ctg-wp-2013-31.pdf [accessed: 21 July 2013].

Polanyi, M. 1958. *Personal Knowledge.* Chicago: University of Chicago Press.

Popper, K. 1967/1983. Knowledge: Subjective versus objective, in *A Pocket Popper*, edited by D. Miller. Oxford: Oxford University Press, 58–77.

Prahalad, C.K. and Bettis, R. 1986. The dominant logic: A new linkage between diversity and performance. *Strategic Management Journal* 7(6), 485–501.

Reed I. 2003. Structural hermeneutics and the possibility of a cultural science. *Yale Journal of Sociology* 3(Fall), 105–11.

Reichertz, J. 2004. Objective hermeneutics and hermeneutic sociology of knowledge, in *Companion to Qualitative Research*, edited by U. Flick, E. von Kardoff and I. Steinke. London: Sage 290–95.

Rennstam, J. 2012. Object-Control: A Study of technologically dense knowledge work. *Organizational Studies* 33(8), 1971–90.

Revilla, E. and Rodríguez, B. 2011. Team vision in product development: How knowledge strategy matters. *Technovation* 31(2–3), 118–27.

Rizzello, S. and Spada, A. 2012. The knowledge-rationality connection in Herbert Simon, in *Handbook of Knowledge and Economics*, edited by R. Arena, A. Festré and L. Lazaric. Cheltenham and Northampton: Edward Elgar, 144–64.

Roberts, J. 2013. Communities as spaces of innovation, in *Innovation in Socio-Cultural Context*, edited by F. Adam and H. Westlund. New York and London: Routledge, 80–100.

Roland Berger (2013a). *Corporate Headquarters.* [Online]. Available at: http://www.rolandberger.com/media/pdf/Roland_Berger_Corporate_Headquarters_Short_version_20130502.pdf [accessed: 4 September 2013].

Roland Berger (2013b). *Frugal Products.* [Online]. Available at: http://www.rolandberger.com/media/pdf/Roland_Berger_Frugal_products_20130212.pdf [accessed: 4 September 2013].

Rose, G. and Tolia-Kelly, D.P. 2012. Visuality/materiality: Introducing a manifesto for practice, in *Visuality/Materiality: Images, Objects and Practices*, edited by G. Rose and D.P. Tolia-Kelly. Farnham and Burlington: Ashgate, 1–11.

Rouse, W.B. and Morris, N.M. 1986. On looking into the black box: Prospects and limits in the search for mental models. *Psychological Bulletin* 100(3), 359–63.

Rüb, S. 2002. *World Works Councils and Other Forms of Global Employee Representation in Transnational Undertakings. A Survey.* Düsseldorf: Hans Böckler Foundation.

Rüb, S., Platzer, H.-W. and Müller, T. 2011. *Transnationale Unternehmensverein-barungen: Zur Neuordnung der Arbeitsbeziehungen in Europa. [Transnational Company Agreements: The Restructering of Labour Relations in Europe].* Berlin: Edition Sigma.

Rutten, R. and Boekema, F. 2012. From learning region to learning in a socio-spatial context. *Regional Studies* 46(8), 981–92.

Rutten, R. and Irawati, D. 2013. Learning in regional networks: The role of social capital, in *Innovation in Socio-Cultural Context*, edited by F. Adam and H. Westlund. New York and London: Routledge, 126–41.

Sacchetti, S. 2009. Introduction, in *Knowledge in the Development of Economies. Institutional Choices under Globalisation*, edited by S. Sacchetti and R. Sugden. Cheltenham and Northampton: Edward Elgar, 3–15.

Sacchetti, S. and Sugden, R. 2009. Positioning order, disorder and creativity in research choices on local development, in *Knowledge in the Development of Economies. Institutional Choices under Globalisation,* edited by S. Sacchetti and R. Sugden. Cheltenham and Northampton: Edward Elgar, 269–88.

Saxenian, A. 2012. The new argonauts, global search and local institution building, in *Innovation, Global Change and Territorial Resilience*, edited by P. Cooke, M.D. Parrilli and J.L. Curbelo. Cheltenham and Northampton: Edward Elgar, 25–42.

Schamp, E.W. and Stamm, A. 2012. New trends in an old sector: Exploring global knowledge and HR management in MNCs and the North–South divide in human capital formation. *Innovation and Development* 2(2), 285–302.

Schamp, E.W. 2012. Constructing a global centre for competence from local knowledge: The case of Pirmasens. *Urbani Izziv* 23(1), 94–103.

Scharpf, F.W. 1997. *Games Real Actors Play.* Boulder: Westview.

Scheler, M. 1925/1960. *Die Wissensformen und die Gesellschaft. [The Shapes of Knowledge and the Society].* Bern and Munich: Francke Publisher.

Schilirò, D. 2010. Investing in knowledge: Knowledge, human capital and institutions for the long run growth, in *Governance of Innovation: Firms, Clusters and Institutions in a Changing Setting*, edited by M.J. Arentsen, W. van Rossum and A.E. Steenge. Cheltenham and Northampton: Edward Elgar, 33–50.

Schiller, D. 2013. *An Institutional Perspective on Production and Upgrading. The Electronics Industry in Hong Kong and the Pearl River Delta.* Stuttgart: Franz Steiner Publisher.

Schlögel, K. 2003. *Im Raume lesen wir die Zeit:* Über *Zivilisationsgeschichte und Geopolitik. [In the Space We Can Read the Time: About History of Civilisation and Geopolitics].* München: Carl Hanser Publisher.

Schmid, S. and Daniel, A. 2011. Headquarters-subsidiary relationships from a social psychological perspective: How perception gaps concerning the subsidiary's role may lead to conflicts, in *Politics and Power in the Multinational Corporation: The Role of Institutions, Interests, and Identities*, edited by C. Dörrenbächer and M. Geppert. Cambridge and New York: Cambridge University Press, 255–80.

Schmitz, H. 2004a. Globalized localities: Introduction, in *Local Enterprises in the Global Economy*, edited by H. Schmitz. Cheltenham and Northampton: Edward Elgar, 1–19.

Schmitz, H. 2004b. *Local Enterprises in the Global Economy*. Cheltenham and Northampton: Edward Elgar.

Schoenberger, E. 1999. The firm in the region and the region in the firm, in *The New Industrial Geography: Regions, Regulations and Institutions*, edited by T.J. Barnes and M.S. Gertler. London and New York: Routledge, 205–24.

Scholz, R. 2013. *Mitbestimmung, Partizipation und Kompetenzentwicklung im Maschinen- und Anlagenbau in ausgewählten Regionen Deutschlands, Schwedens und der Schweiz. [Co-Determination, Participation and Competencies: Mechanical Engineering Industry in Selected Regions of Germany, Sweden and Switzerland]*. Cologne: University of Cologne, Department of Economic and Social Geography, Working Papers 2013–3. [Online]. Available at: http://www.wigeo.uni-koeln.de/11070.html [accessed: 24 September 2013].

Schotter, A. and Beamish, P.W. 2011. Intra-organizational turbulences in multinational corporations, *in Politics and Power in the Multinational Corporation: The Role of Institutions, Interests, and Identities*, edited by C. Dörrenbächer and M. Geppert. Cambridge and New York: Cambridge University Press, 191–230.

Schütz, A. 1943. The problem of rationality in the social world. *Economica, New Series* 10(38), 130–49.

Segrestin, B., Lefebvre, P. and Weil, B. 2002. The role of design regimes in the coordination of competencies and the conditions for inter-firm cooperation. *International Journal Automotive Technology and Management* 2(1), 63–84.

Selwyn, B. 2012a. Beyond firm-centrism: Re-integrating labour and capitalism into global commodity chain analysis. *Journal of Economic Geography* 12(1), 205–26.

Selwyn, B. 2012b. *Workers, State and Development in Brazil: Powers of Labour, Chains of Value*. Manchester and New York: Manchester University Press and Palgrave Macmillan.

Shapin, S. 1998. Placing the view from nowhere: Historical and sociological problems in the location of science. *Transactions of the Institute of British Geographers. New Series* 23(1), 5–12.

Sheppard, E. 2002. The spaces and times of globalization: Place, scale, networks and positionality. *Economic Geography* 78(3), 307–30.

Simon, H. 1957. *Models of Man*. New York and London: Wiley.

Soeffner, H.-G. 2004. *Auslegung des Alltags: Der Alltag der Auslegung. [Interpretation of Everyday Life: Everyday Life of Interpretation]*. Konstanz: UVK Publisher.

Som, O. 2012. Innovation systems without R&D? A retrospective view of five years of 'low-tech' innovation research at Fraunhofer ISI, in *Innovation System Revisited: Experiences from 40 Years of Fraunhofer ISI Research*, edited by Fraunhofer Institute for Systems and Innovation Research ISI. Stuttgart: Fraunhofer Publisher, 43–64.

Som, O. and Jäger, A. 2012. *Qualität auf dem Vormarsch. [Quality in Advance]*. Karlsruhe: Fraunhofer ISI.

Song, Y.I., Lee, D.-H., Lee, Y.-G. and Chung, Y.-C. 2007. Managing uncertainty and ambiguity in frontier R&D projects: A Korean case study. *Journal of Engineering and Technology Management* 24(3), 231–50.

Spencer, G.M. 2011. Local diversity and creative local activity in Canadian city-regions, in *Beyond Territory: Dynamic Geographies of Knowledge Creation, Diffusion, and Innovation*, edited by H. Bathelt, M.P. Feldman and D.F. Kogler. London and New York: Routledge, 46–63.

Srinivas, S. and Sutz, J. 2008. Developing countries and innovation: Searching for a new analytical approach. *Technology in Society* 30(2), 129–40.

Stegmaier, J. 2012. Effects of works councils on firm-provided further training in Germany. *British Journal of Industrial Relations* 50(4), 667–89.

Storper, M. 1995. The resurgence of regional economies, ten years later: The region as a nexus of untraded interdependencies. *European Urban and Regional Studies* 2(3), 191–221.

Storper, M. 2008. Community and economics, in *Organising for Creativity: Community, Economy and Space*, edited by A. Amin and J. Roberts. Oxford: Oxford University Press, 37–68.

Strube, G., Thalemann, S., Wittstruck, B. and Garg, K. 2005. Knowledge sharing in teams of heterogeneous experts, in *Barriers and Biases in Computer-mediated Knowledge Communication*, edited by R. Bromme, F.W. Hesse and H. Spada. New York: Springer, 193–212.

Sturgeon, T. and Florida, R. 2000. Globalisation and Jobs in the Automotive Industry. Final Report to the Alfred P. Sloan Foundation. Cambridge: MITIPC.

Sturgeon, T. 2003. What really goes on in Silicon Valley? Spatial clustering and dispersal in modular production networks. *Journal of Economic Geography* 3(2), 199–225.

Sturgeon, T. and Florida, R.L. 2004. Globalization, deverticalization, and employment in the motor vehicle industry, in *Locating Global Advantage*, edited by M. Kenney and R.L. Florida. Stanford: Stanford University Press, 52–81.

Sturgeon, T., van Biesebroeck, J., Gereffi, G. 2008. Value chains, networks and clusters: Reframing the global automotive industry. *Journal of Economic Geography* 8(3), 297–321.

Sundbo, J. and Toivonen, M. 2011. Introduction, in *User-based Innovation in Services*, edited by J. Sundbo and M. Toivonen. Cheltenham and Northampton: Edward Elgar, 1–21.

Sunley, P., Pinch, S., Reimer, S. and Macmillen, J. 2008. Innovation in a creative production system: The case of design. *Journal of Economic Geography* 8(5), 675–98.

Swyngedouw, E. 1997. Neither global nor local: 'Glocalization' and the politics of scale, in *Spaces of Globalization: Reasserting the Power of the Local*, edited by K. Cox. New York: Guilford Press, 138–66.

Tavares, A. and Young, S. 2006. Sourcing patterns of foreign-owned multinational subsidiaries in Europe. *Regional Studies* 40(6), 583–99.

Taylor, F.W. 1911. *The Principles of Scientific Management*. London: Harper and Brothers.

Teece, D. and Pisano, G. 1994. The dynamic capabilities of firms: An introduction, *Industrial and Corporate Change* 3(3), 537–56.

Teirlinck, P. and Spithoven, A. 2008. The spatial organization of innovation: Open innovation, external knowledge relations and urban structure. *Regional Studies* 42(5), 689–704.

The Economist 2006. *The Battle for Brainpower*. [Online]. Available at: http://www.economist.com/node/7961894 [accessed: 5 September 2013].

Thompson, P. and Harley, B. 2012. Beneath the radar? A critical realist analysis of 'the knowledge economy' and 'shareholder value' as competing discourses. *Organization Studies* 33(10), 1363–81.

Thrift, N. 2000. Pandora's box? Cultural geographies of economies, in *The Oxford Handbook of Economic Geography*, edited by G.L. Clark, M. Feldman and M.S. Gertler. Oxford: Oxford University Press, 689–704.

Thrift, N. 2009. Space: The fundamental stuff of geography, in *Key Concepts in Geography. Second Edition*, edited by N.J. Clifford, S.L. Holloway, S.P. Rice and G. Valentine. Los Angeles, London, New Delhi, Singapore and Washington: Sage, 85–96.

Tödtling, F., Prud'homme van Reine, P. and Dörhöfer, S. 2011. Open innovation and regional culture: Findings from different industrial and regional settings. *European Planning Studies* 19(11), 1885–907.

Tokatli, N. 2013. Toward a better understanding of the apparel industry: A critique of the upgrading literature. *Journal of Economic Geography* 13(6), 993–1011.

Tomiura, E., Ito, B. and Wakasugi, R. 2013. Offshoring and Japanese firms, in *The Oxford Handbook of Offshoring and Global Employment*, edited by A. Bardhan, D.M. Jaffee and C.A. Kroll. New York: Oxford University Press, 229–51.

Torre, A. and Rallet, A. 2005. Proximity and localization. *Regional Studies* 39(1), 47–59.

Törnqvist, G. 2011. *The Geography of Creativity*. Cheltenham and Northampton: Edward Elgar.

Truffer, B. and Coenen, L. 2012. Environmental innovation and sustainability transitions in regional studies. *Regional Studies* 46(1), 1–21.

TRW 2012. *TRW Automotive Profile 2012*. [Online]. Available at: http://www.trw. com/AboutTRW [accessed: 28 May 2013].

TRW 2013a. *TRW Reports First Quarter 2013: Financial Results*. [Online]. Available at: http://trw.mediaroom.com/index.php?s=32950&item=128097 [accessed: 28 May 2013].

TRW 2013b. *Historia. [History]*. [Online]. Available at: http://kariera.trw.pl/ pruszkow/historia [accessed: 28 May 2013].

Ullrich, C. 1999. Deutungsmusteranalyse und diskursives Interview [Patterns of interpretation and discoursive interview]. *Zeitschrift für Soziologie* 28(6), 429–47.

UNCTAD 2012. *Technology and Innovation Report 2012*. New York and Geneva: United Nations Publications.

UNCTAD 2013. *Global Value Chains and Development. Investment and Value Added Trade in the Global Economy*. New York and Geneva: United Nations Publication.

Van den Bossche, P., Gijselaers, W., Segers, M., Woltjer, G. and Kirschner, P. 2011. Team learning: Building shared mental models. *Instructional Science* 39(3), 283–301.

Van Dijk, J. and Bosch, S. 2003. Regional development and the role of company-provided training, in *Knowledge, Learning, and Regional Development*, edited by V. Lo and E. Schamp. Münster: Lit, 83–105.

Van Egeraat, C. and Breathnach, P. 2012. The drivers of transnational subsidiary evolution: The upgrading of process R&D in the Irish pharmaceutical industry. *Regional Studies* 46(9), 1153–67.

Veblen, T. 1898. Why is economics not an evolutionary science? *The Quarterly Journal of Economics* 12(4), 373–97.

Veblen, T. 1910. *The theory of business enterprise*. New York: C. Scribner's sons.

Veltz, P. 1996. *Mondalisation, Villes et Territoires: L' économie d' archipel. [Globalisation, Cities and Territories: The archipelago economy]*. Paris: Presses Universitaires de France.

Volkswagen 2012a. *Volkswagen Group Academy*. [Online]. Available at: http:// www.volkswagenag.com/content/vwcorp/info_center/de/news/2012/08/Volks wagen_Group_Academy.html [accessed: 30 May 2013].

Volkswagen 2012b. *Vocational Training at Volkswagen*. [Online]. Available at: http://annualreport2012.volkswagenag.com/managementreport/value-enhan cingfactors/employees/vocationaltraining.html [accessed: 30 May 2013].

Volkswagen 2012c. *Volkswagen Charter on Labour Relations*. [Online]. Available at: http://nachhaltigkeitsbericht2012.volkswagenag.com/en/society/ employment/charter-on-labour-relations.html?tx_eepcollect_pi1%5Bprozess %5D=add&tx_eepcollect_pi1%5Bpid%5D=137&tx_eepcollect_pi1%5Bctrl %5D=1367038191 [accessed: 16 August 2013].

Volkswagen 2013a. *Media Services*. [Online]. Available at: https://www.volks wagen-media-services.com/medias_publish/ms/content/de/reden/2013/03/14/

Winterkorn-JPK2013-TeilIII.standard.gid-oeffentlichkeit.html [accessed: 30 May 2013].

Volkswagen 2013b. *Volkswagen Academy Enhances Mechatronic Programs to Associate Degree.* [Online]. Available at: http://www.volkswagengroupamerica. com/newsroom/2013/03/01_Volkswagen_Academy_Enhances_Mechatronic_ Programs_to_Associate_Degree.html [accessed: 30 May 2013].

Volkswagen 2013c. *Volkswagen Navarre.* [Online]. Available at: http://www.vw-navarra.es [accessed: 21 July 2013].

Volkswagen 2013d. *Volkswagen Navigator 2013.* [Online]. Available at: http:// www.volkswagenag.com/content/vwcorp/content/en/investor_relations.html [accessed 26 July 2013].

Von Behr, M. 2012. Von der rechnerintegrierten Fertigung zur international verteilten Produktion [From computer integrated production to internationally distributed production], in *Sozialwissenschaftliche Beiträge zur Produktions forschung [Contibutions in Social Sciences to Research about Production]*, edited by H. Hirsch-Kreinsen, G. Lay and J. Abel. Stuttgart: Fraunhofer Publisher, 33–42.

Wagner, G. and Vormbusch, U. 2010. Informal networks as 'global microstructures': The case of German expatriates in Russia. *Critical Perspectives on International Business* 6(4), 216–36.

Wagner, S., Lukassen, P. and Mahlendorf, M.D. 2010. Misused and missed use: Grounded theory and objective hermeneutics as methods for research in marketing. *Industrial Marketing Management* 39(1), 5–15.

Wallerstein, I. 1979. *The Capitalist World-Economy.* Cambridge: Cambridge University Press.

Weber, M. 1920/1986. *Gesammelte Aufsätze zur Religionssoziologie. [Collective Essays about the Sociology of Religion]. Volume 1.* Tübingen: Mohr Publisher.

Weiss, P. 2005. *The Aesthetics of Resistance. Volume 1.* Durham: Duke University Press Books.

Wenger, E. 1998. *Communities of Practice. Learning, Meaning and Identity.* Cambridge: Cambridge University Press.

Werner, M. 2012. Beyond upgrading: Gendered labor and the restructuring of firms in the Dominican Republic. *Economic Geography* 88(4), 403–22.

Westeren, K.I. 2012. Developments in the analysis of the knowledge economy: Introductory comments, in *Foundations of the Knowledge Economy: Innovation, Learning and Clusters*, edited by K.I. Westeren. Cheltenham and Northampton: Edward Elgar, 1–12.

Westlund, H. and Kobayashi, K. 2013. Social capital and sustainable urban-rural relationships in the global knowledge society, in *Social Capital and Rural Development in the Knowledge Society*, edited by H. Westlund and K. Kobayashi. Cheltenham and Northampton: Edward Elgar, 1–17.

Westlund, H. and Li, Y. 2013. Collaboration in innovation systems and the significance of social capital, in *Innovation in Socio-Cultural Context*, edited by F. Adam and H. Westlund. New York and London: Routledge, 126–41.

Wetzstein, S. and Le Herron, R. 2010. Regional economic policy 'in-the-making': Imaginaries, political projects and institutions for Auckland's economic transformation. *Environment and Planning A* 42(8), 1902–24.

Wiesenthal, H. 1995. Konventionelles und unkonventionelles Organisationslernen. [Conventional and unconventional learning of organisations]. *Zeitschrift für Soziologie* 34(2), 137–55.

Will-Zochol, M 2011. *Wissensarbeit in der Automobilindustrie. [Knowledge Work in the Automobile Industry]*. Berlin: Edition Sigma.

Williams, K. and Geppert, M. 2011. Bargaining globalization: Employment relations providing robust 'tool kits' for socio-political strategizing in MNCs in Germany, in *Politics and Power in the Multinational Corporation: The Role of Institutions, Interests, and Identities*, edited by C. Dörrenbächer and M. Geppert. Cambridge and New York: Cambridge University Press, 72–100.

Willke, H. 2001. Die Krisis des Wissens. [The crisis of knowledge]. Österreichische *Zeitschrift für Soziologie* 26(1), 3–26.

Wills, J. 2002. Bargaining for the space to organize in the global economy: A review of the Accor-IUF trade union rights agreement. *Review of International Political Economy* 9(4), 675–700.

Winkler, D. 2013. Services offshoring and the relative demand for white-collar workers in German manufacturing, in *The Oxford Handbook of Offshoring and Global Employment*, edited by A. Bardhan, D.M. Jaffee and C.A. Kroll. New York: Oxford University Press, 72–99.

Winter, J. 2008. Spatial division of competencies and local upgrading in the automotive industry, in *Globalising Worlds and New Economic Configurations*, edited by C. Tamasy and M. Taylor. Aldershot: Ashgate, 113–24.

Winter, J. 2009. *Zwischen Hierarchie und Heterarchie. Kompetenzveränderungen in Tochterbetrieben internationaler Automobilunternehmen am Standort Polen. [Between Hierarchy and Heterarchy. Changing Competencies in Subsidiaries of International Automotive Companies in Poland]*. Münster: Lit.

Witt, U. 1998. Imagination and leadership: The neglected dimension of an evolutionary theory of the firm. *Journal of Economic Behavior and Organization* 35(2), 161–77.

Witt, U., Broekel, T. and Brenner, T. 2012. Knowledge and its economic characteristics: A conceptual clarification, in *Handbook of Knowledge and Economics*, edited by R. Arena, A. Festré and L. Lazaric. Cheltenham and Northampton: Edward Elgar, 369–82.

Womack, J., Jones, D. and Roos, D. 1990. *The Machine that Changed the World*. New York: Free Press.

Wong, A., Tjosvold, D. and Liu, C. 2009. Cross-functional team organizational citizenship behavior in China: Shared vision and goal interdependence among departments. *Journal of Applied Social Psychology* 39(12), 2879–909.

Wyckoff, A.W. 2013. The OECD innovation strategy: Science, technology and innovation indicators and innovation policy, in *Handbook of Innovation*

Indicators and Measurement, edited by F. Gault. Cheltenham and Northampton: Edward Elgar, 301–19.

Yang, D.Y.-R. and Chen, Y.-C. 2013. ODM model and co-evolution in the global notebook PC Industry: Evidence from Taiwan. *Advances in Applied Sociology* 3(1), 69–78.

Zanker, C., Kinkel, S. and Maloca, S. 2013. *Globale Produktion von einer starken Heimatbasis aus. [Global Production with a Strong Home Base].* Karlsruhe: Fraunhofer ISI.

Index

.